Power from the Earth

POWER FROM THE EARTH

*Deep Earth Gas —
Energy for the Future*

Thomas Gold

J. M. Dent & Sons Ltd
London Melbourne

First published 1987
© Thomas Gold 1987

This book is set in 11/13 Linotron Sabon by
Input Typesetting Ltd, London SW19 8DR
Printed in Great Britain by
Mackays of Chatham Ltd for
J. M. Dent & Sons Ltd
Aldine House, 33 Welbeck Street, London W1M 8LX

British Library Cataloguing in Publication Data

Gold, Thomas
    Power from the earth: deep earth gas —
    energy for the future.
    I. Power resources
    I. Title
    333.79   TJ163.2
    ISBN 0–460–04462–1

CONTENTS

List of Illustrations   vi
Preface   vii

Introduction   1
1 The great debate   9
2 Where did the carbon come from?   21
3 The origin of diamonds   38
4 Earthquakes   45
5 Outgassing in solid rock   78
6 Eruptions   92
7 The evidence from helium   104
8 Where are oil and gas found?   123
9 The origin of petroleum   134
10 Conclusions   166

Appendix 1: The carbon isotopes   175
Appendix 2: Speculations about the deep interior of the   185
        Earth
References   189
Glossary   198
Index   203

LIST OF ILLUSTRATIONS

The main structure of the Earth

1a & b The supply of carbon from deep sources in the Earth to the atmosphere
2 Model of a Kimberlite pipe
3 Global map of the occurrence of Kimberlites and diamonds
4 Pressure regime in rock and fluid-filled pore-spaces
5 Fluid-filled domains in a vertical succession of pressure regimes
6 'Overpressure' or 'underpressure' where there is a gas reservoir
7 Drilling with a heavy mud through the critical layer
8 Geologically young faults and fold belts
9 Relationships in the proportions of helium, nitrogen and methane in the Hugoton-Panhandle area
10 Two models to account for the helium-nitrogen-methane relationships in the Hugoton-Panhandle fields
11 Mideastern oil and gas fields
12 Relation between volcanoes, earthquakes and petroleum occurrence in Indonesia and Burma
13 Stability of hydrocarbons at temperatures and pressures in the Earth
14 Density of methane at depth in the Earth
15 Optical activity in different fractions of petroleum, in different temperature reservoirs
16 Methane hydrates on the ocean floor
17 Distribution of the ratio of stable isotopes carbon-13 and carbon-12 in different terrestrial materials
18 Carbon isotope ratios of methane plotted against depth of occurrence
19 Comparison of carbon isotope ratios of methane and co-existing carbon dioxide in ocean floor sediments

PREFACE

My thanks are due to many colleagues and supporters who helped me and encouraged me to do the work presented here. Most help and the guidance through the extensive literature on the subject was provided by Dr Steven Soter, and it was he who tracked down the remarkable earthquake evidence of historical times. Mr Marshall Held was responsible for the collection of many data, especially those relating to the systematic relations between nitrogen, helium and methane.

The Gas Research Institute (Chicago) supported much of the research, and its President, Dr Henry Linden, spurred me on by his interest and enthusiasm. I greatly valued also the encouragement extended by Mr Robert A. Hefner III and Mr Ed Schmidt.

The discovery, in the course of this work, that several Soviet scientists, notably Dr Peter N. Kropotkin, have developed similar views and have collected much evidence to support them, has been a great help. My apologies to them or to other contributors to the subject whose work may not be adequately referenced.

My thanks are also due to my wife for the immense patience and hard work in the preparation of the typescript.

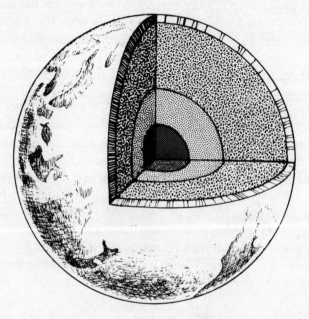

Cut-away diagram showing the main structure of the Earth: a crust of hard, brittle rock; a mantle composed of a pliable hot solid; a liquid core; and a solid inner core.

INTRODUCTION

Much of the fascination of the study of the Earth comes from its fantastic complexity. There are vast numbers of clues from which to make deductions about the many processes that have taken place. Some of these processes have come to be understood, but for many others there are only suggestions. Occasionally a definitive proof is found that turns one of these suggestions into firm knowledge.

A large number of plays have been enacted on this grand stage, but for a majority of them the script has been lost and they will be forgotten for ever; just an occasional accidental mark left here or there may allow some fragment of the action to be reconstructed.

What can one hope to achieve? Can one hope to trace out the major outline of how the Earth formed? How the surface rocks were made? Where the atmosphere and the oceans came from? Can one find out what forces built the mountains, and what forces displaced large blocks of the crust? What is it that causes earthquakes, or the gigantic eruptions that we know occurred in the past?

Although the Earth is 4½ billion years old, it seems that its interior has not yet settled down to a quiescent state. The details of the physical and chemical processes that are taking place down there are not yet known, and are very hard to study theoretically; the experimental approach is also of limited value only, for the processes occur in regimes of temperature, pressure and elapsed time that cannot be modelled correctly in the laboratory. Direct observation is not possible, because dense matter does not allow much information to go through. Much more information reaches us from distant galaxies than from any place beneath the outer few kilometres of the Earth.

This is not to deny that great progress has been made. We have to appreciate the magnitude of the subject, the complexity of the clues, to understand how it is possible for much to have been achieved, and yet for most of the major questions to remain unanswered. But the experimental techniques and the sophistication of the arguments are rapidly improving, and they will advance and reshape our present understanding. Some advances will build on what is now generally believed. Other advances will have to await the wholesale abandonment of present beliefs, to be supplanted by a new edifice of ideas. That is the way in which science has progressed in the past, and most likely it will continue to do so in the future.

The study of the Earth clearly ought to be treated as a single subject, since all the many aspects are interconnected. Yet it is so vast a subject that in practice this has not been possible. Separate disciplines have grown up, and the field has been carved up between them. The geologists, the geochemists, the geophysicists, the cosmochemists, have each singled out some part of the problem to be studied, and they each know that they cannot work independently of experts in the other fields. A lot of detail has been studied in this way, but perhaps a lot of perspective has been lost.

Opinions on major topics in the Earth sciences have often fluctuated between extremes. The timescales of geology were once the subject of major controversies. The modern view, widely held, of the movement of continents and the subduction of the ocean floor, is an example of a recent swing of the pendulum of opinion. Not long ago any such viewpoint was generally regarded as outrageous, while now it is firmly entrenched. Major aspects of the behaviour of the Earth are not yet understood, such as the origin of the forces that push the continents around, and in the face of such uncertainty we cannot feel very secure with explanations of any particular one of the processes going on inside the Earth.

The origin of the variety of carbon deposits that we find in the crust of the Earth is the subject that will concern us here. It is a subject of considerable theoretical interest and practical importance, but it stretches across virtually all the disciplines into which Earth science has been subdivided. Cosmochemistry and the study of the formation processes of the Solar System and of the Earth itself have advanced enough to give us information about the types of material and the forms in which the carbon may have been supplied initially. Geochemistry tells us something about the chemical processing that may have taken place on the surface and in the crust, and geology informs us about the multitude of physical and chemical effects that have shaped the deposits and the settings in which we find them.

In the last eighty years or so a particular viewpoint has grown up, and has been almost universally accepted, that attributes nearly all deposits of unoxidized carbon − such as oil, coal, and natural gas − to a biological origin. Atmospheric carbon dioxide and solar energy, through the process of photosynthesis in plants, gave rise to unoxidized carbon, and this, sequestered and protected from oxidation in the sediments, is then considered to have become the source material for all carbonaceous deposits (i.e. the deposits of largely unoxidized carbon, including the great variety of hydrocarbons, the hydrogen-carbon compounds that make up petroleum and natural gas). Many observations and investigations of recent times do not accord well with this viewpoint, by now conventional, and some dissatisfaction has been noted here and there in the literature. But to replace a viewpoint that stretches across so many disciplines is, of course, very difficult. Can an alternative viewpoint be suggested that will accord better with the vast array of evidence? At what stage does it become worthwhile to rework all the evidence in another framework? If the conventional theory is right, then this would be merely a laborious and useless exercise. If it is wrong, then the longer one delays, the more laborious and the more painful would be the process of re-evaluation.

The alternative viewpoint, that hydrocarbons come from deep in the Earth, from materials that were incorporated as the Earth formed, is clearly the minority viewpoint, but nevertheless one that has had a significant following ever since the subject of the origin of hydrocarbons first came to be debated, more than a hundred years ago. In 1877 Mendeleev, the great Russian chemist who discovered the periodic nature of the chemical elements, published a paper on the subject in which he gave a number of reasons why he considered petroleum to be emanating from deep in the Earth, and not from organic sediments. He saw that the occurrences of petroleum seemed to be controlled more by the large-scale features of the crust of the Earth, like the mountain ranges and the great valleys, than by the details of sedimentary deposits laid down over the ages. A substantial number of investigators have held such a viewpoint over the years since then up to the present and have advanced many detailed arguments to support it. If we argue in these pages once more in favour of such a viewpoint, it is only because much modern information seems to have strengthened it greatly, and at the same time weakened the case for the biological origin.

Great quantities of carbon must have emerged from the interior of the Earth, because the surface and the sediments contain enormously more carbon than could have been derived from the erosion and weathering of the rocks that originally formed the crust of the Earth, the "basement rocks". If hydrocarbons from the deep Earth were responsible wholly or in part for this large supply of carbon, then we might be dealing with far larger quantities than were ever contemplated to be present in biological deposits. In that case the problem of the origin of the hydrocarbons which we use as fuels would not only be one of great scientific interest, but also one of great practical significance. The estimate of the quantities yet to be discovered, the possible locations and the techniques and strategies of search, may all be greatly affected.

In recent years the judgement that oil and gas are running out has had a profound influence on the world economy and on world politics. The dramatic rise in the price of these fuels, the shift in the wealth of nations and the political tensions that centre around the access to the great oil-fields of the world – all these have resulted from the *prediction* of a shortage and not from a shortage itself. Yet despite the immense and frightening importance that this prediction has had, it has gone almost unchallenged and hardly debated. Is it really so secure a prediction that there is no further need to look into the scientific basis?

Few of those who made the prediction would claim that it is more than their best judgment, made from a limited amount of information. But a best judgment is not necessarily a secure judgment.

Oil is rarely found at the deeper levels; the cut-off is generally around 15,000 feet. Exploration to date over most of the Earth at these shallower levels has been extensive, though by no means exhaustive yet. Quite likely there will be further important discoveries on the scale of the North Slope of Alaska or the new Mexican fields. With a different theory of the origin of oil, quite a number of new areas might come under consideration that seemed to have no potential before. One may postpone the oil-crunch by a few decades, but it would be hard to argue that we are not nearing the end. At the present rate of use there may be enough for quite a few decades, but not for centuries.

For gas the outlook is completely different. Much gas is found unassociated with oil, and the deeper levels, of which our knowledge and experience are very limited, contain large quantities of it, at least in some areas. One can certainly not claim that the Earth has been exhaustively explored for deep gas, nor can one claim to understand the patterns of distribution empirically. If any judgment has been offered that gas in the depth range from 15,000 to 30,000 feet is in short supply on the Earth, such a judgment is certainly quite insecure. There are vast areas in all parts of the world where very little is known about the subsurface at these depths and where drilling may very well result in high success rates. Fifteen years ago virtually nothing was known about gas at such depths, and few suspected that much could be found there. Now great quantities have been discovered in this domain in the few areas that have been tested, mainly in Oklahoma and on the north shore of the Gulf of Mexico, and very productive wells are flowing. In the Deep Anadarko Basin of Oklahoma the success rate for deep wells drilled in recent years is said to be 70 percent, the highest in oil and gas exploration history. How many other such areas will be found?

If we were convinced that all hydrocarbons on the Earth, including gas at these levels, are due entirely to biogenic deposits, then perhaps we could make a judgment about the amounts of gas that might be down there. Even for that the limits would be very wide, and the amounts that could be there are very large. But if supplies of gas to these levels could have come from below, from source materials incorporated at much deeper levels in the Earth when it formed, the prognosis might be quite different. The quantities temporarily arrested on the way up in this deeper but accessible depth range might then be enormous. The limits are then set only by the volumes of porous rock that may exist there, under conditions where porosity may be created or held open by the supply of a very high pressure gas from below. No organic sediments need be involved. No rock that acts as a gas-tight seal needs to have been positioned over a gas reservoir at these deeper levels, since any rock will create an adequate barrier at such depths and hinder the upward migration. The absence of these two requirements for a gas reservoir increases the possibly productive volumes by factors of hundreds. From a scientific as well as from a practical point of view, it is well worth investigating this possibility.

There are many other reasons for wanting to understand the outgassing processes of the Earth. The origin of many mineral deposits appears to be related to the upward flow of liquids and gases from deep levels. The methods of prospecting for minerals may therefore be improved by a better understanding of these flow processes.

Another range of phenomena that appear to be linked to the upward flow of gases through the crust is that of earthquakes and giant eruptions. Fluids are involved in triggering the violent shearing of rock-masses that causes the great earthquakes, and gas emissions are sometimes observed accompanying them or as precursory events.

Huge quantities of gas are involved in some volcanic eruptions, including those eruptions which have devastated vast areas in prehistoric times. Perhaps these processes can be understood and measurements devised to predict them.

Changes in the composition of the atmosphere are being investigated intensively at present. There are man-made changes in carbon dioxide content that could have harmful effects. There are the natural compositional changes that seem to cause great waves of climatic changes, including the ice ages. Do emissions of gases from deep levels play a part in these effects? Do they occur all the time, continually and gradually, so as to go unobserved? Or are there occasional big releases? Are the man-made effects significant against the background of the natural phenomena? Clearly we would like to know these things before deciding on major modifications of global industrial policy. In any case we need to understand how safe or precarious our existence on this planet really is, and the composition of our frail atmosphere must give us much cause for concern.

The long-term movement of major pieces of the Earth's crust is under investigation and is thought by many to account for a wide range of observations and effects. What drives such processes? Where is the engine and what is its fuel? Density changes of large masses in crust or mantle must be involved, but what causes them? Are chemical differentiation processes at work in the mantle? Are fluids that percolate through the deep rocks responsible for chemical processes that generate density differences, thereby driving a flow? Is chemical energy so released a significant item in the heat budget of the Earth? These are among the many further problems that give reasons for studying the outgassing processes that bring volatile substances from great depth to the surface.

In these pages we shall be discussing quite a number of items that may be relevant to these outgassing processes of the Earth as well as to the question of the origin of all the surface carbon. We shall make suggestions as to how the diverse data that we now have may fit together in a coherent picture, and we are quite aware that not all of these are going to be the final answers. But the final answers in this field are not yet at hand, whether within the conventional set of ideas or within any novel picture.

What we can do at this stage is to enlarge the discussion to encompass modern knowledge, to include the understanding we now have of the chemistry of the early Solar System, of the formation processes of the planets, and the manner in which carbon was incorporated into them. We can also consider now, within this new framework, which carbon compounds could be available in the pressure and temperature regimes we know to exist in the Earth, and what fates they would suffer.

In this field, as in any field of science, all theories and interpretations should be reconsidered whenever new information becomes available. In the field of the origin of terrestrial hydrocarbons, a large change in the basic information certainly has occurred since the adoption of the theories of bio-origin.

I became interested in this subject in the mid 1950s while concerning myself with the large-scale mechanical properties of the body of the Earth (Gold, 1955). At that time I started to think about the manner in which fluids would penetrate through solid rock, upwards if they were lighter, and downwards if they were heavier than the rock material. Fred Hoyle (1955) in his book *Frontiers of Astronomy* gives a good discussion in a section entitled "Gold's Pore Theory" which already foreshadows several of the points discussed in the present book, including the possibility of a deep origin of petroleum.

The greatly renewed interest in the planets that resulted during the 1970s from the space-probe investigations caused me to think once more about the origin of the planetary bodies and the various volatile substances of their atmospheres. Of course the Earth had to be thought of as one of those planets, and the one we should surely understand the best. It was in the course of studying again, at that time, the information, not just about the hydrocarbons, but about the origin of the element carbon itself on the surface of the Earth, that I became surprised how little was known and how insecure all the information really was. There was no discussion at all about how the element carbon came to be concentrated on the surface of our planet, and in the context of the other planets it seemed unavoidable to me to consider hydrocarbons as the most likely source material on the Earth also. But to challenge the established viewpoint that the terrestrial hydrocarbons were of biological origin required the pursuit of many detailed arguments. After all, if a single one of the many quoted indications of a biological origin could not be challenged, the established theory would stand. An established theory always enjoys a large advantage, and the heresy has to gain much more than an equal position of strength before it is taken seriously.

I
THE GREAT DEBATE

"It is remarkable that in spite of its widespread occurrence, its great economic importance, and the immense amount of fine research devoted to it, there perhaps still remain more uncertainties concerning the origin of petroleum than that of any other commonly occurring natural substance."

H. D. Hedberg, 1964

"Actually it cannot be too strongly emphasized that petroleum does not present the composition picture expected from modified biogenic products, and all the arguments from the constituents of ancient oils fit equally well, or better, with the conception of a primordial hydrocarbon mixture to which bio-products have been added."

Sir Robert Robinson, 1963
President, Royal Society, London

"The capital fact to note is that petroleum was born in the depths of the Earth, and that it is only there that we must seek its origin."

D. Mendeleev, 1877

The origin of petroleum has been a subject of many intense and heated discussions, ever since this black fluid was first discovered to be present in large quantities in the pore spaces of many rocks. Is it something brought in from space when the Earth was formed? Or is it a fluid concentrated from huge amounts of vegetation and animal remains that have been buried in the sediments over hundreds of millions of years?

Arguments have been advanced for each viewpoint, and although they appear to conflict with each other, each line of argument sounds strangely convincing. The answer has to be a rather subtle one, for on each side there are points that cannot be ignored.

In favour of a *biogenic* origin of petroleum, the following four observations have been advanced:

(1) Petroleum contains groups of molecules which are clearly

identified as the break-down products of complex, but common, organic molecules that occur in plants, and that could not have been built up in a non-biological process.

(2) Petroleum frequently shows the phenomenon of optical activity, i.e. a rotation of the plane of polarization when polarized light is passed through it. This implies that molecules which can have either a right-handed or a left-handed symmetry are not equally represented, but that one symmetry is preferred. This is normally a characteristic of biological materials.

(3) Some petroleums show a clear preference for molecules with an odd number of carbon atoms over those with an even number. Such an odd-even effect can be understood as arising from the break-down of a class of molecules that are common in biological substances, and may be difficult to account for in other ways.

(4) Petroleum is mostly found in sedimentary deposits and only rarely in the primary rocks of the crust below; even among the sediments it has been said to favour those that are geologically young. In many cases such sediments appeared to be rich in biological materials that could have been the source material for the petroleum deposit.

On the other side of the argument, in favour of an origin from deeply buried materials incorporated in the Earth when it formed, the following observations have been cited:

(1) Petroleum and methane are found frequently in geographic patterns of long lines or arcs, which are related more to deep-seated large-scale structural features of the crust, than to the smaller scale patchwork of the sedimentary deposits.

(2) Hydrocarbon-rich areas tend to be hydrocarbon-rich at many different levels, corresponding to quite different geological epochs, and extending down to the crystalline basement that underlies the sediments. An invasion of an area from below could better account for this than the chance of successive deposition.

(3) Some petroleums from deeper and hotter levels lack the biological evidence almost completely. Optical activity and the odd-even carbon number effect are sometimes totally absent, and it would be difficult to suppose that such a thorough destruction of the biological molecules had occurred as would be required to account for this.

(4) Methane is found in many locations where a biogenic origin is improbable: in great ocean rifts in the absence of any substantial sediments; in fissures in igneous and metamorphic rocks, even at

great depth; in active volcanic regions, even where there is a minimum of sediments; and there are massive amounts of methane hydrates (methane-water ice combinations), in permafrost and ocean deposits, often with little underlying sediment.

(5) The hydrocarbon deposits of a large area often show common chemical features, quite independent of the varied composition or the geological ages of the formations in which they are found. Such chemical "signatures" may be seen in the abundance ratios of some minor constituents such as traces of certain metals that are carried in petroleum; or a common tendency may be seen in the ratio of isotopes of some elements, or in the abundance ratio of some of the different molecules that make up petroleum. Thus a chemical analysis of a sample of petroleum could often allow the general area of its origin to be identified, even though quite different formations in that area may be producing petroleum. For example a crude oil from anywhere in the Middle East can be distinguished from an oil originating in any part of South America, or from the oils of West Africa.

(6) The regional association of hydrocarbons with the inert gas helium, and a higher level of natural helium seepage in petroleum-bearing regions, has no explanation in the theories of biological origin.

We shall return to each of these points with a more detailed discussion.

The Earth contains many other deposits whose origin is not yet clear. Metal ores and many other commercially valuable minerals have been laid down, sometimes in remarkably high concentrations, by processes that are still uncertain. In many cases a variety of chemical and physical processes could have led to the same end result; even a most detailed analysis of the end product does not give sufficient evidence about the process that put it there. The situation is no different with petroleum. The fact that biogenic materials can suffer a conversion process to a mix of hydrocarbon molecules which we call petroleum is no proof that all petroleum arises in such a fashion; equally the demonstration that petroleum can arise without any biological materials would be no proof of an *abiogenic* origin of all petroleum found on the Earth.

At this point it is helpful to look at the evolution of ideas in this field and the state of knowledge that led to them. At a time when the biological origin theory of hydrocarbons came to be largely adopted – perhaps about eighty to a hundred years ago – the relevant background information accepted was very different from what we know today. It was then thought that the Earth had been formed as a molten sphere of rock and had slowly cooled to produce a solid crust. Hydrocarbons clearly could not have survived such a history, and the possibility of a hydrocarbon source material imbedded deep in the Earth could not even be considered.

Hydrocarbons were regarded at that time as a typical product of biological processes, and the idea of hydrocarbons derived without the intervention of biology seemed remote. This viewpoint was so prevalent that when methane was discovered on Jupiter, papers were published that this surely proved an abundance of life out there. Today we recognize that hydrocarbons are common in the cosmos, and in our planetary system much more of the element carbon has been observed in the form of hydrocarbon molecules than in any other form.

The estimates that were made eighty years ago of the quantities of oil and gas to be found on the Earth predicted only a very small fraction of what has in fact been discovered, certainly less than 1%. If one considers the huge amounts of methane now thought to reside on the ocean floor and in permafrost in combination with water ice (methane hydrates), then the early estimates would be for less than one ten thousandth part; and what we know today may still be only a small fraction of what is really there. The biogenic theory therefore started out with a very modest requirement so far as quantities were concerned; somewhat unusual circumstances of deposition of organic materials and their subsequent modification could be invoked to account for the deposits.

Now, with enormously larger quantities to account for, the conviction seems to have grown up, without any other justification, that such processes were not unusual but the norm, that very large amounts of organic material got buried without being first oxidized, and that the application of heat and pressure and catalytic action was often available to produce the large quantities of hydrocarbon fluids and gases which are found. In favour of this it was argued that very large amounts of unoxidized carbon in the form of dense, largely insoluble compounds (named "kerogen"), distributed in the rocks (indeed not only in the sediments, but also in crystalline rocks), are of biological origin; this kerogen was then considered to be the source material for the fluid hydrocarbons. It is difficult here to separate the evidence from the interpretation: the more kerogen and other carbonaceous material that is found, the more it is taken to prove the importance of the biological processes. If, on the other hand, unoxidized carbon compounds, including hydrocarbons, are thought to be a significant complement of the Earth in the first place, then the vast majority of these carbon deposits could be interpreted as deriving from this primordial material, and not from surface biology.

While the origin of kerogen and carbonaceous materials is uncertain, the origin of recognizable fossils is not. If one could establish that sediments rich in carbonaceous materials, such as that dense agglomerate of carbon compounds called kerogen, petroleum and natural gas, were generally also particularly rich in fossils, this would strengthen the case for a biogenic origin. But this is not so. Much petroleum is found in sediments that contain no fossils; natural gas is found in huge quantities at deep levels in some areas in layers that are totally free from fossils. Carbonaceous materials, including coal, graphite and tar, exist in many areas of basement rock, in igneous rock or in ancient granite, in many parts of the world (Powers *et al.*, 1932). Very large coal deposits are known that are free from fossils.

Shale, a dense deposit that sometimes contains fossils, often has a high carbon content and is frequently regarded as the "source rock" of oil fields in the vicinity. But such an association could also be interpreted as due to both the presence of shale and the position of the oil field in the same region of upwelling hydrocarbons, both having been supplied by the same stream. Shale would in fact act as a particularly effective molecular sieve and absorber, extracting heavy hydrocarbon molecules from any such stream. Where there are fossils, the total amount of carbon deposited by them, even if they had all been buried without being oxidized, would in all cases only be a small fraction of the carbon in the deposits.

The quantitative aspects cannot settle the question, since we have no basis for judging how much more unoxidized carbon was deposited together with the quantity originally contained in all the fossils. The chemical similarity of oil to its supposed "source rock", which is often cited to prove the biological origin of oil, would be meaningful only if one could prove the biological origin of the carbon in this source rock; if both had come from the same underlying supply, the similarity would be assured anyway.

The belief that carbonaceous material was always of biological origin caused many deductions to be made on the basis of the deposits that were found. If such a deposit was found in very ancient

sediments that had previously been thought to pre-date the emerg-
ence of life, this was regarded as a proof that life already existed
earlier. If carbonaceous material was found in igneous or metamorphic rocks, it was taken as indicating that a subducted sediment lay
beneath. If it was discovered that in the same region vertically
stacked sedimentary layers of quite different ages were all rich in
carbonaceous material, the conclusion was that the area had a
particular propensity for rich vegetation in different geological
epochs, even if no direct reason for this could be found. But if
carbonaceous materials could be delivered directly from below, all
these points would have to be argued afresh.

More recent evidence should cause us to re-examine a great many
of the interpretations of the past. We now know for sure that the
Earth was built up from solids, that it was never all molten, that
the mantle of the Earth is not thoroughly mixed, and that volatiles
do come up from great depth. We have good reason now for thinking
that hydrocarbons were included in the forming Earth and that their
outgassing became the major source of the great carbon excess of
the surface and the sediments. But even after this new information
had become available, hydrocarbon outgassing from great depth was
still not generally considered a possibility. If any carbon came up
from below, it was thought that it had to be in the form of carbon
dioxide.

There were two reasons for this. One was that methane and other
hydrocarbons were considered to be so unstable against thermal
dissociation that they could not survive the high temperatures which
rule at depths below about 10 or 15 kilometres; they would have
been decomposed below into immobile graphite and free hydrogen
(which would probably combine with oxygen in the rocks and form
water). The new evidence does not support this. High temperatures
go with high pressures in the depths of the Earth, and it has recently
been recognized that hydrocarbon molecules are greatly stabilized
by pressure. A depth of origin of hydrocarbon molecules between
150 and 300 km is now considered possible (Chekaliuk, 1976);
indeed hydrocarbons are found in diamonds, which are thought to
derive from such depths.

Secondly, it was thought that even if any hydrocarbons could
survive the temperatures, they could not survive being oxidized with
the oxygen available in the deeper rocks. Calculations were made
for the equilibrium chemistry that would apply if small amounts
of methane were to find their chemical equilibrium in the oxygen

environment provided by the deep rocks. There was a large measure of uncertainty about the type of rock to be considered, and its oxygen "fugacity" (a measure of the available oxygen), but it appeared that the more oxidizing varieties would have caused complete destruction of the hydrocarbons. On the other hand, if one assumed the least oxidizing of these possible rocks, the hydrocarbons would essentially have been completely preserved (Arculus and Delano, 1980). It was thought that the calculations of the chemical equilibrium could not be decisive until one knew more about the oxygen availability of the deep rocks.

Fortunately, however, we do not have to await this information. The mechanical process of outgassing through the solid rocks changes the discussion in a radical way. Gases that are in low abundance cannot make their way up; they are permanently imprisoned in the high-pressure rock. It is only where gases are abundant enough to be able to create fractures through which they can move that an effective outgassing can take place. But in such fractures there will be a lot of gas in contact with a limited area of rock surface. Oxygen from the rock can be supplied only by diffusion through the solid, and the amount that can reach the crack surfaces is limited. Once this is used up in the oxidation of methane, further methane (and other hydrocarbons) can be conducted through the fractures without chemical depletion. The chemical equilibrium to be discussed for such a case is not that of a large amount of rock with a little methane, but rather of a large amount of methane with a small amount of rock.

The situation is of course completely different if the pathways go through liquid rock, known as magma. There each bubble of gas making its way through has fresh access to oxygen contained in the magma. The conventional calculations would apply in the case of a large amount of magma and a trickle of hydrocarbon fluids making their way through it. This is the situation in volcanic areas, where the carbon is indeed emitted principally as carbon dioxide (although some methane is frequently present also). Even if methane were the source material, it would be largely oxidized to carbon dioxide and water in such circumstances. Similarly, inclusions in an igneous rock may contain carbon chiefly as carbon dioxide, which would have been placed there when the rock was molten, while cracks in the same rock may now carry methane.

If methane and other hydrocarbons largely survive in the temperature-pressure regime down to depths of several hundred kilometres, and if hydrocarbon fluids that are abundant enough to force their way up through solid rock largely survive the ascent without being destroyed, a deep origin of methane and petroleum has to be considered as a serious possibility.

Advocates of the abiogenic theory

Mendeleev's paper on the origin of petroleum is still well worth reading. He was followed by a succession of investigators, mostly, but not entirely, in Russia who contributed many important considerations in favour of the abiogenic theory. W. Sokoloff, in 1889, published a paper on the "cosmic origin of bitumina", bitumina being a term used to describe the whole range of carbonaceous substances from petroleum to pitch and tar. He related the terrestrial bituminous substances to the meteorites, knowing already then that some meteorites contained bituminous materials. He could find no relationship between the occurrence of fossils and terrestrial hydrocarbon deposits and stated that "porosity is the sole circumstance which relates to the accumulation of bituminous substances". He stressed that oil and tar are found in crystalline basement rocks, as for example in the gneiss of Sweden, as well as in basalt, as at the base of Mount Etna. He argued against an origin of hydrocarbons from a reaction of metal carbides and water, when a simpler explanation in terms of a formation from carbon and hydrogen seemed perfectly satisfactory. He seems already then to have been aware that the Earth may have formed from meteorite-like materials and in circumstances where hydrogen was plentiful.

V. I. Vernadsky (1933) gave reasons why he considered that with depth the availability of oxygen in the rocks is rapidly reduced, and under those circumstances, with the increased pressure at depth, he considered that hydrocarbon compounds would be stable, and largely replace carbon dioxide as the chief carbon-bearing fluids.

N. A. Kudryavtsev (1959) argued that no petroleum has ever been made in the laboratory from genuine plant material. He noted and gave many examples of substantial and often commercial quantities of petroleum being found in crystalline or metamorphic basements, or in sediments directly overlying such basement. He cited cases in Kansas, California, Western Venezuela and Morocco. He also pointed out that oil pools in sedimentary strata are frequently related to fractures in the basement directly below. The Lost Soldier oil field in Wyoming has oil pools, he stated, at every horizon of the geological section, starting with the Cambrian sandstone immediately overlying the ancient crystalline basement, and continuing up to the upper Cretaceous deposits. A flow of oil, though not of commercial significance, was also obtained from the basement itself. He noted that hydrocarbon gases are not rare in the igneous and metamorphic rocks of the Canadian Shield. Petroleum in Precambrian gneiss is encountered in the wells on the eastern shore of Lake Baikal. He realized that petroleum is present, albeit in small quantity only, in all horizons downwards below *any* accumulation, totally independent of the composition and conditions of formation of the rocks which make up these horizons, and that commercial accumulations form simply in those rocks in which the permeable zones are overlain by impermeable. This statement has since become known as "Kudryavtsev's rule", and many examples of it have been noted (Kropotkin and Valyaev, 1984).

Kudryavtsev also introduced into the argument a number of other relevant observations. Columns of flames have been seen during the eruptions of some volcanoes, and sometimes reach 500 metres in height, such as the Merapi eruption in Sumatra in 1932. He observed that the loss of a great quantity of gas (mostly methane) liberated from mud volcanoes during their violent eruptions should have exhausted any underlying pool, even if it was the most prolific ever known, and the eruptions should therefore have ceased long ago, even if they were due to every gas deposit that could have existed in the area. Geological investigations have shown that mud volcanoes on the Kerch Peninsula (Southern USSR) which are active now were already erupting as early as middle-Miocene (around 10 million years ago). Several thousand eruptions are needed to explain the quantities of mud deposited. No gas reservoir big enough to explain that fact is known anywhere.

Kudryavtsev also observed that the chemical make-up of the water in mud volcanoes is sometimes very strange and suggested quite a different origin from locally available water. Sometimes, for example, it contains quantities of iodine, bromine and boron that exceed the content of these elements in sea water a hundredfold and that could not have been derived from the local sediments. He also noted that mud volcanoes are often associated with lava volcanoes, and the typical relationship was that where they are close, the mud volcanoes emit incombustible gases, but where they are further away from the lava volcanoes, they usually emit methane. He observed that there are numerous occurrences of hydrocarbons in the kimberlite pipes (the channels of eruptions of gases from great depths) in South Africa and in the Soviet Union. He knew of the presence of oil in the Kola Peninsula in basement rock, and also in a ring-shaped patch in Central Sweden (the Siljan Ring, which we shall discuss later in these pages). He discussed the enormous quantity of hydrocarbons in the Athabasca tar sands in Canada, and the vast amount of source rocks with organic content which would have been necessary in the conventional biological explanation to account for this, when in fact no source rocks at all have been found.

Beskrovny, in 1968, discussed details of the formation of petroleum hydrocarbons by nonbiological means, and he noted, as Anders, Hayatsu and Studier also noted in 1973, that of all the possible molecules that could be assembled from a given number of hydrogen and carbon atoms, the same sub-set seems to be singled out in artificial oil production as in natural petroleum.

V. B. Porfir'ev (1974) dealt with all aspects of the debate. He argued that so-called source rocks have no identification that proves their hydrocarbons to be primarily biogenic. The hypothesis often advanced, that the transport and deposition of oil from the supposed source material to the final reservoir was done by solution in gas, is not tenable within the organic theories, since the quantity of gas needed for this transport would be orders of magnitude larger than could be made from the suggested source materials.

In 1969, Levin wrote that the presence of abiogenic, organic compounds in meteorites makes it practically certain that these compounds were present also in the matter from which the Earth has accumulated. Judging from the composition of a class of meteorites called carbonaceous chondrites, the general volume of these primary organic compounds brought to the Earth could have been several times larger than the entire volume of ocean water.

A. I. Kravtsov (1975) wrote about the inorganic generation of oil and presented much detailed observational material. He discussed the magnitude of natural seepage of methane and showed that this is far larger than anything that could be supplied by the kind of gas fields that are known. If the volcanic gases in the Kurile Islands, for example, are typical of the gases emitted in these islands over the period of volcanic activity there, the amount of methane brought up there would far exceed the largest estimates of present-day total world reserves. He demonstrated the experimental synthesis of hydrocarbons from carbon dioxide and hydrogen, and he gave

numerous examples of "Kudryavtsev's rule", namely that in petroliferous regions, deposits of gas and oil span all horizons in the
vertical direction and can be found down to the base of the sedimentary strata.

P. N. Kropotkin and B. M. Valyaev (1976) developed many
aspects of a theory of deep-seated, inorganic origin of hydrocarbons.
They concluded that petroleum deposits were formed where pressure
conditions permitted the condensation of heavy hydrocarbons, trans-
ferred from great depths by rapidly rising streams of compressed
gases. They distinguished between the ascent of such fluids in
volcanic regions, which would favour the decomposition of hydrocarbons and the formation of carbon dioxide and water, and "cool"
regions where hydrocarbons would be preserved, and could accumulate in the alluvial cover and in highly fractured beds, depending on
the existence of adequate reservoirs and covers. According to these
authors, "Vertical migration of hydrocarbons from levels far below
formations rich in biogenic organic matter, which have been
considered the source material for the oil, can be demonstrated in a
majority of deposits".

In more recent years E. E. Voronoy (1984) has discussed the
mobilization of carbonaceous material – kerogen and coal – by a
stream of hydrocarbon fluids from deep sources, and he regards this
as the main process of oil formation. According to him, "The essence
of processes of hydrogenized mobilization of the kerogen-coal matter
in rock is that the latter acts as a natural solvent". He notes that
methane can cause the deposition of carbon in kerogen-coal matter
and can be a hydrogen donor, causing the hydrogenization of such
matter. "In strata where organic matter was absent or insufficient,"
he says, "gas or gas condensate deposits were formed. In deposits
with high concentrations of organic matter, petroleum hydrocarbons
were generated, producing oil accumulations." He quotes many
cases of coal-type substances deposited along volcanic or other
vertical pathways.

There were several voices also outside the Soviet Union that spoke
in favour of a largely abiogenic origin of oil and gas. Most notable
among them was Sir Robert Robinson (1963, 1966), who, like
Mendeleev, can be considered among the most distinguished chem-
ists of his day. His investigations of the detailed nature of petroleum
led him to conclude that it was generally much more hydrogen-rich
and much less oxidized than would be expected if it had been
generated from plant material. It appeared to him that an origin of

a hydrogen-rich primordial material, later contaminated in one way or another by biological substances, would accord best with the composition of natural petroleum. The details he gives of the abundance ratios of particular molecules in natural petroleum fit remarkably well with similar abundance ratios in artificially produced oils.

Sylvester-Bradley (1964, 1972) has discussed the origin of carbon on the Earth, the presence of hydrocarbons in the meteorites, and the likelihood that petroleum is derived in major part from such material. He combines another subject in this discussion, namely that of the origin of life. If hydrocarbon fluids have been streaming up through the crust, then, he argues, they would have been able to provide the energy source for simple forms of life. Possibly it was there, and not on the surface, that the chemical evolution took place that resulted in the simplest living organisms. He notes that all crude oils now contain live active bacteria, able to withstand high pressures and relatively high temperatures, and that the products of their activity contribute to the total content of the oil and could account for all its biological properties.

Very recently P. N. Kropotkin (1985) has given a historical account of the evolution of ideas, and a general survey of the reasons for believing that hydrocarbon fluids – gas and oil – have come up from deep levels. He cites many examples where oil and gas are found in locations within the basement rocks or just above them, and where no organic deposits can be invoked to account for such deposits. He also gives numerous examples where "Kudryavtsev's rule" is satisfied in a striking way.

These and other advocates of an abiogenic or mixed origin of hydrocarbons on the Earth have contributed many of the detailed arguments that we shall discuss further in these pages.

2

WHERE DID THE CARBON COME FROM?

The Earth appears to be 4½ billion years old. Most of our geological information concerns only the last 450 million years, or 10 percent of that period. The earlier times have not yet been deciphered very well, and we do not know such important items as the quantity of water in the oceans, the quantity and composition of the atmosphere, the forms and the quantity of life, or the size, shape and distribution of continents and water. It is natural, therefore, that the science of geology tried to make itself as independent of these unknowns as possible. The feeling was that it was better to try to explain all that is now observed in terms of processes that could be shown to have occurred in these last 450 million years, rather than to invoke processes from earlier times, including the time of formation of the Earth as a whole. "Let us explain as much as we can by the working and reworking of the crust, without taking recourse to the unknown that went before" – that seems to have been the attitude for the most part, and, of course, a perfectly reasonable attitude it was. However, it must not be taken too far, because there can be processes that do not have their origin in the present crust, and they must not be excluded from discussion. The origin of most of the light elements that formed the volatiles on the surface of the Earth is a major case in point.

The origins of the water of the oceans, the nitrogen of the atmosphere and the carbon which forms so large a part of the biosphere and the sedimentary rocks, must all be sought in some outgassing process that allowed fluids to come up from the deep interior, to be reworked in many subtle ways on or near the surface. The erosion of the original rocks from which the sediments were produced would not have provided these substances. It is clear that here we must appeal to the largely unknown deeper levels, and indeed to the formation processes of the Earth itself.

Much progress has been made in the last twenty years in understanding the chemistry of the cosmos, the formation processes that made the Solar System, the chemical composition of the other planetary bodies, and of the meteorites, which we recognize as planetary debris left over from the time of planetary formation. This recent knowledge has not yet been fully integrated into the geological discussion. The analysis of the reworking of the crust did not need the new insights, but the enrichment of the crust with the volatiles and the carbon is another matter. Here it would clearly help to know how these elements were incorporated into the Earth and what has happened to them since.

The terrestrial or inner planets, Mercury, Venus, the Earth and Mars, had quite a different process of formation from that of the outer planets, Jupiter, Saturn, Uranus and Neptune (about Pluto we know very little). The inner group, so called because they are closer to the Sun, accumulated mostly or entirely from solids, from rock-forming substances and metals, with very little or no contribution from liquids or gases. The outer planets may have some rock-forming minerals in them also, but the bulk of their mass was contributed in the form of ices of the various combinations of hydrogen, carbon, nitrogen and oxygen, and for Jupiter and Saturn there was an additional accumulation, accounting in fact for the bulk of their masses, of gaseous hydrogen and helium. The numerous satellites of these bodies are made from mixes of the rock-forming elements and the ices in various proportions. Also the comets, which cruise on elongated orbits in the Solar System, are composed of a mix of the light-element ices and rock particles.

The inner planets, including our Earth, either acquired no gaseous materials in the first place, or if they did, they must have lost these very completely at an early stage. The evidence for this comes from the very low abundance of the gas neon and other non-radiogenic noble gases – in other words, those isotopes not derived from a radioactive decay process in the case of helium, neon, argon, krypton and xenon. Different components of a gaseous mix could not have become completely separated from each other while all were in gaseous form. A primordial gas mix added to the Earth would have given it an atmosphere containing very much more of these noble gases, in proportion to the nitrogen, than is now the case. The hydrogen and helium of such a primordial mix could well have evaporated from the Earth into space, because these molecules and atoms are light enough to be thermally flung off from the outer atmosphere. Chemically reactive elements could have been removed from the atmosphere by some surface chemical processes; but the noble gases that are both heavy enough to be retained gravitationally, and chemically inert so that they could not have been taken into the crust – those would still have to be present.

In the atmosphere of the planet Venus a small but significant fraction of such gases has been detected and it may be that some solar mix contributed to its atmosphere. For our Earth, the case is clear: we own virtually no solar mix gas. All the volatiles that we have here must have condensed from the original gas into solids, or undergone some chemical reaction that made them end up in a solid phase. In the form of solids, this component can then become separated from the rest of the gaseous mix. The water and the nitrogen of the Earth must therefore have been tied down as solids at some stage, and in this way have become separated from the neon and the other inert gases that were present. Similarly, the large amount of carbon now in the sediments must in the first place have been supplied to the Earth in solid form, and it must then have been turned into a gas in order to be able to concentrate at the surface and result in the identified carbon deposits in the sediments that have derived from an atmospheric gas. The nature of the solids that brought in all these substances that turned into volatiles, including the carbon, is clearly of concern to us here, and the evidence for this can come both from a knowledge of the Solar System chemistry and from clues about the outgassing processes that we can observe on the Earth.

How large are the quantities of fluids that have made their way up through the crust? The erosion of the rocks that led to the production of the sediments contributed only a minor fraction of all the volatile materials we now find in the oceans and atmosphere, or of the carbon concentrated near the surface or in the sediments. Instead of expressing the quantities in terms of the total number of kilograms for the entire Earth, we prefer to express them in terms of kilograms for each square centimetre of the global surface area. (The total surface area of the Earth is 5.1×10^{18} cm^2.) If the ocean water covered the entire Earth with uniform depth, then this depth would be approximately 3 kilometres. The amount of water thus averages to 3,00 kilograms per square centimetre. There is approximately 1 kilogram of nitrogen per square centimetre, and smaller amounts of various other gases. The total amount of carbon that must have come up to produce all the carbon enrichment of the surface and the sediments seems to be about 20 kilograms per square centimetre, and it may be a great deal more, depending on the amount that we do not know about, because it may have been pulled under in descending sediments to levels that are outside our reach.

Most of this large amount of carbon is in the sediments, in the form of carbonate rocks composed of calcium and magnesium carbonates $CaCO_3$ and $CaMg(CO_3)_2$. Almost all these carbonates were derived from atmospheric carbon dioxide dissolved in water and precipitated by reaction with calcium and magnesium. The amount of carbon in the atmosphere, almost all in the form of carbon dioxide, amounts to a little over one ten thousandth of a kilogram per square centimetre – $1.2 \times 10^{-4} kg/cm^2$. (Carbon dioxide is a little more than 0.03 percent by volume, or 0.045 percent by weight, of air). This is of course only a very small fraction of the total "carbon excess" of the crust; approximately one hundred thousandth part, according to these figures. The amount of carbon dissolved as carbonate in the oceans is considerably more, namely about 0.0078 kilograms per square centimetre. The amount of carbon in unoxidized form in the sediments and in the igneous rocks of the crust has been judged to be a quarter or a fifth of the amount of carbon in the form of carbonates, which would make it between 4 and 5 kilograms per square centimetre. These figures of course are not very secure because we may be unable to see subducted sediments that were produced in the earlier geological epochs.

What is the explanation usually offered for the origin of this carbon? Geological texts do not treat this problem very much. Rubey (1951) was the first to concern himself seriously with this subject, and he derived figures for the quantitative aspects of outgassing. Various discussions exist in the literature since then about the forms that the outgassing has taken, but very little has been said about the suggested source material for the various volatiles. For carbon, it seems to have been generally assumed that carbon dioxide has come up from some unknown source material at depth. One reason for this assumption was that volcanoes emit large amounts of carbon dioxide, and so it seemed natural to assume that a source of it existed below.

As we have already discussed, a deep source of methane would equally result in carbon dioxide as the dominant carbon gas from volcanoes, because in bubbling through most lavas at a high temperature and low pressure near the surface the methane would be largely converted to carbon dioxide. Equally, bubbles that froze and trapped gases in volcanic rock also contain much carbon dioxide, and active volcanoes and such entrapped gases were the easiest locations for sampling gases that had come up. Gases that percolate through cooler regions, through solid rock, that come up through cracks and fissures, are harder to locate and sample, and the evidence from these is largely ignored. Where concentrated methane was seen, it was automatically assumed to be of biological origin, and therefore it was not put into the discussion of a primary outgassing process. Perhaps the most vital outgassing information was left out for this reason.

An approximation of the "conventional" viewpoint is shown in Figure 1a, b. An unknown primary material gives off carbon dioxide, which comes out largely through volcanic regions. It contributes to the atmospheric carbon dioxide more or less continuously, and atmospheric carbon dioxide is continuously removed, largely by precipitation in the oceans into carbonates. All the time, some atmospheric carbon dioxide is converted into plant material, having been stripped of its oxygen by the process of photosynthesis. The vast majority of the "organic" carbon so produced is again turned into carbon dioxide in the course of decay of plant material, and returned to the atmosphere. Some of it is turned into methane, which also gets into the atmosphere and is there oxidized in a few years, also to carbon dioxide. A small fraction of the plant material at any one time escapes this degradation and is buried in circumstances where it cannot be oxidized, and is eventually trapped in this form in deep sediments. It is this fraction that is thought to produce most of the forms of unoxidized carbon, like methane, petroleum, coal and the diffusely distributed, unoxidized carbon compounds found in the rocks and generally referred to as "kerogen".

If we think that a primary material is likely to have outgassed methane and other hydrocarbons in the first place, then of course the question must arise whether this has contributed directly to the forms of unoxidized carbon found in the crustal rocks. We would then not have to attribute all this carbon to a cycle where it first was oxidized, then reduced by the process of photosynthesis and then buried. Figure 1b shows this scheme, and the possibility of a direct contribution of unoxidized carbon from the primordial source-materials to the deposits we now find.

In what form is carbon in the other bodies of the Solar System? Carbon is the fourth most abundant element in the Solar mix, after hydrogen, helium and oxygen. The low abundance in the primary rocks of the crust is therefore in a sense an anomaly that needs an explanation. In other bodies of the Solar System we see various

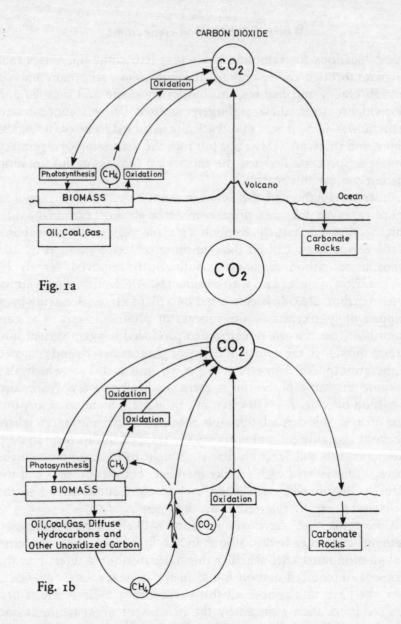

Figs 1a and b: Diagram representing the supply of carbon from deep sources in the Earth to the atmosphere, and its subsequent deposition in the sediments.

CO_2 (carbon dioxide) was thought to be the primary gas responsible (Fig. 1a); CH_4 (methane) is another possibility (Fig. 1b). Carbon would also be oxidized to CO_2 in the shallow ground or in the atmosphere. Atmospheric CO_2 leads to the deposition of the carbonate rocks and also, through the process of photosynthesis, to the deposition of reduced carbon, including substances that could turn to hydrocarbons. If methane and other hydrocarbons come up from below, these may become the main source of the deposits of unoxidized carbon in the ground, including natural gas and petroleum.

degrees of abundance of carbon; the greatest quantity is in the massive outer planets and their satellites. Jupiter, Saturn, Uranus and Neptune have large admixtures of carbon, and they have it in their extensive atmospheres, chiefly in the form of hydrocarbons, mainly methane. Titan, the satellite of Saturn which has the distinction of being a satellite with a substantial atmosphere, shows the carbon there in the form of methane and ethane (CH_4 and C_2H_6). The atmospheric temperature at this great distance from the Sun is of course very low, and it brings the methane–ethane mix into the regime where it can exist both as a vapour and as a liquid. It seems that in the nitrogen atmosphere of Titan, methane and ethane play a similar role to that of water in the terrestrial atmosphere. Clouds are predominantly droplets of these hydrocarbons. Probably it rains methane and ethane, and the surface temperature probably allows methane-ethane liquids to make rivers and lakes, and perhaps even oceans. Methane-ethane ice may exist in the polar regions. (This interesting speculation is due to V. Eshelmann, who conducted studies with the U.S. space instrumented vehicles that investigated the outer planets.)

The asteroids, that enormous swarm of minor planetary bodies that cruise between Mars and Jupiter, also seem to have hydrocarbons on their surfaces (and quite likely also in their interiors). These small bodies cannot retain any atmosphere and would therefore lose any gaseous hydrocarbons. Very dark surfaces and details of the reflection spectrum give the strong suggestion that very heavy hydrocarbons – perhaps tar-like substances – are prominent there.

Venus, the sister planet to the Earth in the sense of being a neighbour with almost the same mass and mean density, has an atmosphere mainly composed of carbon dioxide. This atmosphere is about a hundred times more massive than the Earth's, and it may represent an amount of carbon quite similar to that which seems to have come into the Earth's atmosphere (20 kilogràms of carbon per square centimetre is the figure we have mentioned, and that would make 73 kilograms of carbon dioxide per square centimetre, or 73 atmospheres).

We have no idea yet whether any hydrocarbons exist on Venus, or what the source of the atmospheric carbon dioxide has been there. For Venus, as for the Earth, an outgassing process from initially solid materials must be responsible for the bulk of this atmosphere.

The small planet Mars has only a very tenuous atmosphere with a surface pressure of less than 1 percent of that of the Earth. It also is composed chiefly of carbon dioxide. Again we have no direct evidence for the nature of the source material, but here too the absence of a Solar mix of gases implies outgassing from solids.

We see that the terrestrial planets have the carbon predominantly in oxidized form in their atmospheres, while the outer planets have the carbon in unoxidized form, namely as hydrocarbons. Do we have to ascribe this difference to a different source material for the inner from that for the outer planets? Or is there an explanation in terms of chemical processes that occurred after the formation of these bodies?

For the Earth we know that unoxidized carbon would readily be oxidized in the atmosphere. Any methane, for example, that is put into the atmosphere will be turned into carbon dioxide in a time that is estimated as between 4 and 7 years. But here, in the present geological epoch at least, the atmospheric oxygen for this process is due to biology. Without the process of photosynthesis in plants, the Earth's atmosphere would only have very small amounts of free oxygen. What then is the situation on Venus and Mars? Would hydrocarbons there have been converted to carbon dioxide also, or do we have to suppose that carbon dioxide was the primary outgassing product?

The inner planets develop oxidizing conditions in their atmospheres and on their surfaces, even in the absence of biology. The reason is that sunlight will dissociate some water molecules, and the very light hydrogen so produced will escape into space. This leaves a supply of oxygen which mixes back down into the atmosphere and is available for the oxidation of atmospheric gases or for an extra oxidation of the surface rocks. On Mars it has been possible to observe that the surface rocks are indeed in a very highly oxidized condition. Partial melting in the rocky planets tends to bring the more highly oxidized rocks towards the surface, because they tend to be less dense, and this provides another reason why the rocky inner planets have developed the strongly oxidizing conditions we now find. This means that we can make no deductions about the original supply of carbon there, since it would have ended up as carbon dioxide in either case. Biology on the Earth only helps this process along, but it would have happened without biology just the same.

Why is all the carbon dioxide on Venus in the atmosphere, when a comparable amount on the Earth has been laid down in the rocks?

This difference is accounted for by the presence of liquid water on the Earth, but not on Venus. The bulk of the carbon dioxide that the Earth evolved was precipitated in the oceans and laid down as carbonate rock. Without liquid water the Earth also would have a massive atmosphere dominated by carbon dioxide. On Mars the temperature conditions may have resulted in a much smaller amount of outgassing, and the cold surface makes it possible that solid carbon dioxide has been deposited on the poles. It has also been suggested that liquid water in the soil at equatorial latitudes may exist on Mars in sufficient quantity to have precipitated carbonates. In any case Mars is a very different body from the Earth, and there would be no reason to expect a similar amount of outgassing to have occurred there. The highly oxidized state of the surface rocks makes it clear that we could not expect unoxidized carbon to survive in its atmosphere.

The meteorites provide much evidence about the conditions in the early Solar System. They represent a debris left over from the many and varied processes that took place around the time of formation of the planets. The inner planets formed from solids which had condensed out of a disk-shaped gaseous cloud. Various accumulation and collision processes seem to have produced the original condensate, resulting in different classes of material. The inner planets formed from these materials by an accumulation process which almost completely exhausted the initial supply. The amount of meteoritic material that has survived without being swept up by one or another planet is only a very small fraction of what was once there to build the planets, and it is saved from collision with them by being on orbits that only very rarely cross those of the planets. The different types of meteorites appear to represent materials that were abundant in the formation stage, but the relative quantities of the different types that now hit the Earth may fail to represent the early abundances correctly. A great deal of chance is involved in the deflection into Earth-crossing orbits, and the relative abundances of the different types might well change substantially over time. Still, in spite of this shortcoming, the meteorites provide many clues.

It is clear that among the different types of meteorites we see some that condensed at a fairly high temperature, at which iron and other metals, and the metal oxides that form rocks, would be in solid form. Much of this material seems indeed to have been molten at one time in its accumulation history, probably by being heated by intense early radioactivity in small bodies which were frequently

again destroyed by collisions. This type of high-temperature condensate seems to have provided the bulk of the material from which
Mercury, Venus and the Earth formed. The mean density of these
bodies, the internal density distribution of the Earth, and what
evidence we have of the internal chemical compostion, agree with
that conclusion. If, however, the high-temperature condensate had
been the only source material for these bodies, they would contain
much less of the volatile substances that we now find on the Earth
and on Venus in their surfaces or above them (Mercury is too small
and too close to the Sun to retain any atmosphere, and we know
little or nothing of outgassing processes there).

Some addition of more volatile materials must have taken place
to give rise to the production of the water for the oceans, the
nitrogen for the atmosphere, and the gases or liquids that brought
up carbon to produce this large carbon enrichment of the sedimentary rocks. Yet, as we have said, at the formation of the Earth this
material cannot have been in gaseous form. We therefore need to
look for a material which could have been in solid form then, so as
to be separated from neon and other noble gases, but which later,
in the temperature-pressure circumstances that became established
at some depth in the Earth, would outgas the required volatiles and
dispatch them towards the surface. The requirement is for a low
temperature condensate, solid at the temperatures experienced by
Earth-crossing particles at some phase of the build-up process of the
Earth, but capable of putting out the required gases or liquids after
the incorporation into the Earth. Do we recognize any materials in
the Solar System that would satisfy these requirements?

The class of meteorites called carbonaceous chondrites fit this
requirement very well (Wilkening, 1978). They are indeed a low-
temperature condensate and emit many volatiles when heated. They
contain water, bound to the rocky materials in the form of water
of hydration. They contain carbon, sometimes as much as 5 percent
by weight, which is largely in unoxidized form and includes a range
of hydrocarbon molecules. The rocky material shows the more
refractory elements in much the same ratios as those in which they
occur in the Sun, and one therefore refers to these meteorites as the
most primitive ones – the ones in which the least chemical processing
seems to have taken place. They look more or less like the mix that
you would expect to condense at a low temperature, perhaps 300°K
or less, from a gas cloud of roughly solar composition.

How abundant was this material, and was it really this, or some other material that we have not yet seen, that brought the volatiles to the terrestrial planets?

As we have already noted, we are really not in a position to judge quantitative matters from the meteorites. What falls on the Earth in sufficiently recent times to be identifiable as a meteorite has been subject to chance perturbations, and the statistical information must be very poor. But for what it is worth, the carbonaceous chondrites are indeed thought to be a very abundant variety, possibly the most abundant one in space. They are, however, very much more friable than most other meteorites, and therefore they break up readily in their passage through the atmosphere. Also, any surviving pieces that lie on the ground are destroyed more quickly by weathering, and this again diminishes the chances of finding this type. Among the found meteorites, they are therefore comparatively rare.

There are other ways of making some inferences about the quantities of such materials in the Solar System. Dust which floats in the upper atmosphere, and is thought to be due predominantly to infall from space, has been collected and carefully analysed. A large proportion of the grains show essentially a carbonaceous chondrite composition. Since the bulk of the space dust in our neighbourhood seems to have been shed by comets, we can conclude that the rocky material in comets is also predominantly of this type.

The comets live further out in the Solar System and are only occasionally deflected into the inner, warmer regions. In addition to this carbonaceous chondrite-like rocky material, the comets contain water ice, and the various other ices that can be made from the combination of hydrogen, carbon, nitrogen and oxygen. Out there, where it is cold, such ices could condense and would be retained in solid form. Halley's comet, carefully investigated by space probes in 1986, was found to emit hydrocarbon gases, and the surface of its core was seen to be "black as pitch" (or some other solid carbonaceous material), most likely because that is what it was made of.

Like the comets, the satellites of the outer planets are also composed of mixes of rocks and ices in various proportions, and it is thought likely that all these bodies represent the primordial mix that formed from solids, at progressively lower temperatures the further out from the Sun they condensed.

There is another point to remember: the objects that we now pick up as meteorites have been small chunks, cruising on orbits in the Solar System for 4½ billion years. When they formed from a gaseous cloud, that cloud contained initially all the material that it took to

form the planets. The solids that formed from this cloud were therefore surrounded by a gas many orders of magnitude more dense
than the very tenuous gas at present in the spaces of the Solar
System. Many substances would have been in solid form on these
same meteorites in the circumstances of their formation, but would
have gradually evaporated from them in the high vacuum which they
experienced later. We can only judge from the surviving chemical
composition which types of molecules would have been associated
with them in the early phase. Since the carbonaceous chondrites
now contain hydrocarbon molecules, but predominantly only those
that have a very low volatility, one would certainly guess that at
earlier times more volatile hydrocarbons would have been associated
with them.

Can we identify carbonaceous chondrite material as a component
of the Earth? Some investigators think so, because they recognize
the abundances of certain elements in ratios similar to those found
in the carbonaceous chondrites. The platinum group of elements
(platinum, iridium, rhodium, osmium) has been shown to have abundance ratios in the crust of the Earth quite similar to those in these
meteorites. Furthermore, isotope data for neon, argon and xenon
trapped in trace amounts in the crust of the Earth show a closer
correspondence to these gases in carbonaceous chondrites than to
the trace gases in any other meteoritic materials. Crustal material
seems to have been derived to a significant extent from carbonaceous
chondrite stuff, while the major mass of the Earth comes from a
high-temperature condensate, more similar to a mix of the refractory
meteorites.

Can we understand why, in the formation of the Earth, at least
two distinct source materials should have been involved, and why the
volatile-rich component should have provided at least a sprinkling in
the late stages of the process?

Most of the formation process of the Earth and the other terrestrial
planets occurred in a disk-shaped cloud of grains and gas. So long
as a substantial fraction of the material was still in the form of this
cloud and not yet in the final bodies that were to form from it, the
density in the cloud was so high that it would prevent completely
any motion of particles on irregular orbits. Even when as little as
only 5 percent of the planetary mass remained in the cloud, a grain
would still suffer sufficiently many collisions with other grains to
establish a common stream. This situation enforces essentially a
motion on almost circular orbits in a common plane.

"Wild" orbits – orbits that cross this plane or that are severely elliptical – would be quickly eliminated by collision or friction with the bulk of the cloud material. This means that the forming bodies would collect most of their material in an orderly fashion from the condensates that formed in a ring around their orbit. Later, when space was cleaned up by the accumulation of material into the final bodies, these orbital restrictions must have disappeared. By the time that only 2 or 3 percent of the mass remained in the cloud, it began to be possible for chunks to cruise on very eccentric and very inclined orbits. It is clear also that at this stage the existence of the already massive planetary bodies will have frequently generated gravitational perturbations to produce inclined and eccentric orbits. The last few percent of material to be swept up is therefore likely to have been a most disorderly set.

At this stage the Earth may have acquired some material that had initially condensed much further out but been perturbed inwards, and also material that had condensed much closer to the Sun and suffered an outward perturbation. This means that the outer region of the Earth's mantle – that is, the stuff underneath the crust – is likely to have been bombarded with quite diverse materials. Chunks of volatile-rich material may have fallen in, or layers of small volatile-rich grains may have simmered down onto the surface at that stage in the formation. Major impacts would have made a patchwork out of any neatly layered structure which the infall of small grains would have produced. We would end up, before the formation of the crust, with a chemically and structurally uneven outer mantle, full of craters of all sizes and looking much like the lunar surface. There is now considerable evidence that shows the chemical inhomogeneity of the outer mantle, and even the frequent occurrence of circular features underneath the crust.

We shall later discuss the mechanics of outgassing and shall see that the uneven infall – the chemical patchwork of the outer mantle – is an essential ingredient. If the total material that formed the Earth had been well-mixed before the formation, then even if it contained the same amount of the volatile substances, they could never have come to the surface. In a mixed Earth containing a small proportion of volatiles, these would remain permanently imprisoned, and no significant surface concentration of water, nitrogen or carbon would have taken place.

But if this low-temperature condensate was incorporated in a very uneven distribution in the Earth, principally at late stages in its formation, then we may well expect the subsequent outgassing process to show also a very uneven distribution. Where the concentration was high, gases and liquids may accumulate and force their way up through the crust. But this is an "all-or-nothing" phenomenon. It is only when the fluids can build up pressures sufficient to crack the rock and create pathways through which they can stream that significant amounts would reach the surface. Diffusion of the molecules in the unbroken solid rock is so slow a process that, even in the long spans of geological time, only distances of a few hundred metres could be traversed. Diffusion may be important for the collection of gases from their points of origin into pore spaces within distances of metres, and it may be important for traversing impermeable strata that are only a few metres thick, but for the rest a bulk motion in pore spaces is required. Those fluid-filled pore spaces may of course be large ones, held open by magma through which gases can move, but they may also be small cracks, held open, or even broken open, by the pressure of the volatiles.

In what form would carbon be liberated from carbonaceous chondrite material? To discuss this we first have to see in what form the carbon is in this material, and then what the chemical changes might be that could convert some fraction into liquids or gases.

Up to 5 percent of the mass of some carbonaceous chondrites is carbon. Most of this is in unoxidized form, and only a few percent of it is in the form of carbonates, i.e. the combination of carbon dioxide with metal oxides. The unoxidized carbon is present mainly in dense, aromatic and insoluble compounds, but a certain fraction of more volatile compounds is also there, with a clear preference for the types of molecules that also occur in terrestrial petroleum. Anders *et al.*, (1973) have demonstrated that this array of molecules follows closely the particular pattern obtained by catalytic reactions of carbon monoxide and hydrogen. One can therefore speculate that such reactions took place in the solar nebula, and were in fact responsible for producing the solid compounds of carbon that were incorporated both in the meteorites and in the forming Earth.

These speculations of Anders and others are based on a great amount of detailed evidence. Abundance patterns of the different molecules, covering a great range in carbon number, are almost identical between meteorite samples, samples created in the laboratory by the catalytic reaction of carbon monoxide and hydrogen (called the Fischer-Tropsch process), and indeed terrestrial oils as well. When, for the larger molecules, tens of thousands of structural variants would be possible, it is very significant that the Fischer-Tropsch process seems to single out just the very small number of them that occur both in the meteorites and in terrestrial petroleum, as Anders and colleagues demonstrated so clearly. Straight-chain hydrocarbons are most prominent, followed by slightly branched variants. The light, hydrogen-rich hydrocarbon molecules are present in low abundance only, but these are the most volatile, and have probably been lost during the life of the meteorite. The processes that gave the observed molecules would have provided a large abundance of them at the time of condensation of these materials. What was incorporated in the forming Earth presumably contained a much larger proportion of the light molecules than the present-day meteorites.

Carbon monoxide and hydrogen, together with grains of various metal oxides, are likely to have been a constituent of the solar nebula. Thermodynamic calculations of the equilibrium of carbon, oxygen and hydrogen mixes, in roughly solar proportions, show that carbon monoxide would be the chief form of carbon at high temperatures, but should transform to methane below 600°K (Anders *et al.*, 1973). Harold Urey was the first to discuss the problem of the formation of solids in the presence of grains that could act as catalysts, and he suggested that tarry compounds could be formed in this way and become the chief source of carbon on the terrestrial planets. All the evidence that has been obtained since he made this suggestion, in 1953, greatly strengthens his case.

Could these hydrocarbon materials survive the accretion process onto the Earth? After all, if we are considering a late stage of the process, the Earth was already very massive, and the infall speeds almost as high as they would be today. Doesn't this mean that the incoming material would be heated so much on impact that all these molecules would be dissociated?

Luckily, this is not so. Most of the accretion occurred in the form of small grains falling in, and even a very tenuous atmosphere would be sufficient to arrest them in their infall, without heating them very much. (This is true also of the grains that come from space into the outer atmosphere at the present time.) This material would pack down in layers in the forming Earth, but a neatly layered deposition would then be greatly disturbed by occasional impacts of major pieces. Arcs of circles of all sizes of volatile-rich materials would be generated by such impacts, and we may wonder whether we can still identify some of these in the present-day crust in the form of island arcs, of lines of volcanoes, or of certain fluid-derived deposits.

When this material has been covered over, and is at a depth of more than a hundred kilometres, it will be heated, predominantly by the effect of the radioactive minerals in the rocks. Today the temperature at a hundred kilometres is thought to be between 1100 and 1700°C, and at three hundred kilometres between 1700 and 2400°C. What volatile carbon compounds can exist at these temperatures? At one atmosphere pressure, on the surface of the Earth, none of the hydrocarbon compounds would survive such temperatures. It is quite a different story, though, at the pressures ruling in the Earth. The 100-300 km depth range has pressures between about 30,000 and 100,000 atmospheres, and most of the carbon molecules are greatly stabilized by high pressures. (We shall return to this point later.) Present estimates are that methane is largely stable at these temperatures and pressures, though it may partly dissociate into carbon and hydrogen. Many heavier hydrocarbon molecules are also stable.

At these high temperatures and pressures chemical modifications between the different hydrocarbon molecules will occur freely. For a given mean ratio of hydrogen to carbon, and in the presence of a particular set of catalytic surfaces, a definite statistical distribution of the molecular species would result. Unless the hydrogen content is very low, methane is likely to be the dominant member among the mobile fraction. A lot of carbon in immobile form would also be produced. It is methane, with an admixture of heavier hydrocarbon molecules, that would be given off by the material in these circumstances, and it is this mix that would start its upward migration towards the surface.

If carbonates are present in the original mix, then it is possible that carbon dioxide also would be generated. The evidence from the meteorites, and the cosmochemical considerations, suggest that this would be a minor contribution only. If carbon dioxide as well as methane exist at these deep levels – and there is evidence for this – then it may also be that the carbon dioxide resulted from the oxidation of some of the methane, rather than being a primary product from a source material.

If now, in the present state of knowledge, we were to believe that all hydrocarbons on the Earth derived from plant material, through the process of photosynthesis, we would have to argue what seems a difficult case. While hydrocarbons are the most common form of carbon among the planets, we would have to argue that none of the plentiful hydrocarbons found on the Earth are from a similar source.

The Earth, we would have to say, derived all its large supply of carbon from an unknown source material that produced carbon dioxide, a material that is not prominently represented in meteorites or expected from cosmochemical information. The Earth then produced its complement of hydrocarbons from this by a process unique to this planet, namely the biological process of photosynthesis. It does not sound like a very good case!

What can be said about the quantitative aspects? Could the Earth contain enough of this material to account for the required 20 kilograms per square centimetre to have been delivered to the surface?

Let us make a simple calculation for this. Suppose that in the depth range between 100 and 300 kilometres we have a patchwork in which the carbonaceous chondrite material comprises 20% on an average. In this material, carbon amounts to 5%. This means, on an average, each square centimetre column through the 200 kilometre layer would contain 1% of carbon (5% of 20%), which would translate into 660 kilograms per square centimetre. If one-thirtieth of this had been mobilized and reached the outer crust, it would suffice to account for all the carbon of the carbonate sediments and the sediments of unoxidized carbon. Of course the proportion of carbonaceous chondrite type of material may have been very much larger, and the producing layer much thicker. The fraction that needs to have been mobilized would then be much smaller. All one can really say at this stage is that there is no quantitative problem. Volatile-rich material of sufficient quantity to have supplied the water of the oceans could quite easily have supplied the quantity of hydrocarbons for all the surface carbon.

The next steps in this discussion must be to see whether methane from such deep levels can reach the surface, and what chemical changes it might undergo on the way up. We shall also want to see whether any direct evidence for these processes can be obtained.

In this the remarkable story of the diamonds will help us a great deal, since these gemstones are a direct product of carbon, treated with temperatures and pressures that rule in the depth range of 150 to 300 kilometres. Even though only small quantities of diamonds are found near the surface, the fact that they exist at all tells us that high concentrations of carbon occur at these deep levels. Had the Earth not accreted a material rich in unoxidized carbon, it would be most unlikely that centimetre-sized diamonds could have formed anywhere on it.

3

THE ORIGIN OF DIAMONDS

Where can the source material for all the surface carbon be situated? The continental crust of 30-60 kilometre thickness and the oceanic crust of 10-20 kilometre thickness cannot contain it. These layers were implaced as hot plastic or liquid rock, extruded from the mantle below. Not much volatile material can originate in these layers after their extrusion, and so we must look to the mantle below as the ultimate source. Outgassing, the transport of volatiles from the mantle to the surface, may of course be associated with the extrusion of the crust, but it may also happen independently in unrelated locations. There is good evidence for both processes. What information do we have then about the circumstances in the outer mantle, and in particular about its content of carbon?

Diamonds and the evidence of the great eruptions that have brought them to the surface provide the most direct information. This arises because diamonds can only be formed in the depth range of 100 to 300 kilometres, and they bring with them detailed information about the circumstances there.

Diamond is a form of very pure carbon, assembled in the most tightly bound crystalline configuration known. Chemical theory and the experience in making artificial diamonds show that high pressures of the order of 45 kilobars are needed to produce this dense crystal. Such pressures are found in the Earth only at a depth of 150 kilometres or more, and it is somewhere at such depths that natural diamonds must have been formed. The temperature there exceeds 1000°C.

A diamond placed in an oven and subjected to such high temperatures will turn to graphite, the soft form of carbon, which is stable at low pressures. If a diamond were slowly brought up from the high-pressure region in which it originated, so that at each level it adopted the temperature that is normally there, it would not survive the trip to the surface, but instead it would turn to graphite on the way. This would occur whether the diamond were brought up very gradually by an uplift of a region over geological times, or more rapidly by entrainment in the eruption of upwelling magma. In either case the diamond would disintegrate while traversing the temperature-pressure regime in which it is unstable.

How then can a diamond ever get to the surface? It can only survive if it is cooled so fast that there is not enough time for the crystal structure to come apart. At the surface temperature the diamond is metastable, meaning that this is not its equilibrium configuration, but that it would take a very long time – certainly more than 100 million years – for it to revert to the equilibrium, which of course is graphite. Diamonds are not forever!

The only way in which sufficiently rapid cooling could have been accomplished is if the diamond was brought up by a gas at a very high speed and therefore cooled in the expanding gas. But what kind of an eruption would that be? Obviously we are not dealing with the ordinarily known volcanic processes, for they deliver hot magmas at the surface, in which a diamond could not survive.

The geological situation in which diamonds occur shows that an unusual eruption certainly was involved. Although many diamonds are found dispersed in river gravels, the only concentrated deposits are in the rare "pseudovolcanic" structures called "kimberlite pipes". These are deep, vertical shafts, usually filled with a mixture of rock types, including the diamond-bearing rock called kimberlite (Fig. 2). The pipes are named after the first one to be discovered near the South African town of Kimberley. Probably the dispersed diamonds that are found outside kimberlite pipes trace back to such a pipe in the neighbourhood, which in many cases may be covered by later sediments, and for that reason sometimes difficult to discover. Most of the known kimberlite pipes are in South Africa and Siberia, but there are some in North America, in Australia, and probably also in Brazil, where they may be well hidden under younger sediments (Fig. 3).

Near the surface, a typical kimberlite pipe is a funnel-shaped structure, a few hundred metres in diameter. It narrows with depth, becoming a fissure that presumably extends all the way through the crust into the upper mantle, to the level where the diamonds were formed. Of course no one has traced a kimberlite pipe to these depths; only the top few kilometres have been mined for diamonds. Aside from the diamonds, the kimberlite pipes also contain other high-pressure minerals, including peridotite, which is believed to be one of the principal constituents of the mantle. According to Kennedy and Nordly (1968), "The diamond pipes serve as a window that gives us a look into the Earth. There is probably no other group

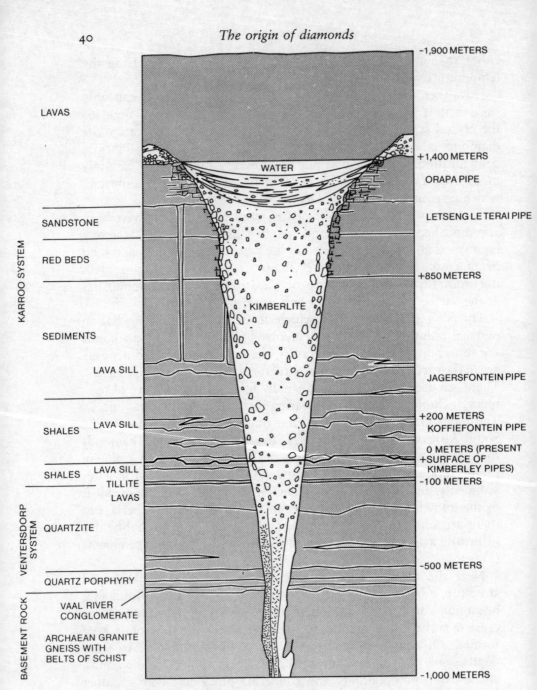

Fig. 2: Model of a Kimberlite pipe (from K. G. Cox, 1978).
The model is based on several South African pipes which have been exposed at various levels by erosion, not only on the one near the town of Kimberley, which gave this formation its name. The model was devised by J. B. Hawthorne of DeBeers Consolidated Mines, Ltd.

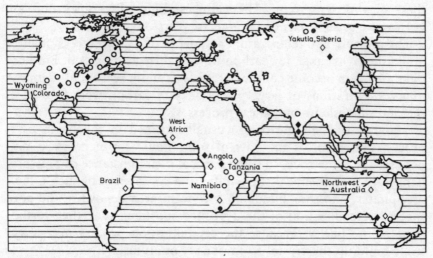

Fig. 3: Global map of the occurrence of Kimberlites and diamonds (from J. D. Pasteris, 1983).
The solid diamond shapes represent locations where Kimberlite eruptions are known to have occurred, and where diamonds are present. The solid circles represent non-diamond-bearing Kimberlite eruptions. The open diamonds represent locations where diamonds are found, but where the site of the eruption has not been identified; the open circles represent the presence of Kimberlite rock without diamonds, and without identified eruption site.

of rocks that originated from even remotely as great a depth as have these."

But in addition to mantle rocks, the pipes also contain an admixture of shallow crustal rocks from all the formations traversed by the eruption on its way to the surface. We have therefore a juxtaposition of fragments, some of high density, which have come up from a depth of 200 kilometres, and others of ordinary surface materials, which have sometimes dropped back a large distance into the open hole. It clearly looks like the result of some extremely violent eruption, but it is reported that rarely if ever has lava been found, either frozen within a kimberlite pipe, or associated with it. For these reasons it is supposed that the kimberlite pipes were produced by an eruption driven essentially by gas, which was so violent that it carried up fragments from the mantle. This then accounts not only for the form and contents of the kimberlite pipes, but also for the rapid cooling that preserved the diamonds.

Fortunately, events of this kind are extremely rare. Although we might like to find more diamonds, we would certainly not like the massive atmospheric pollution that must have been associated with these gigantic eruptions. The youngest kimberlite pipes that are known are several tens of millions of years old.

The pressure and temperature at a depth of 150 or 200 kilometres are in the right range for carbon to crystallize as diamond. But how did the carbon become concentrated? We cannot reasonably suppose that concentrations of large pieces of pure carbon were formed in the outer mantle by biological processes.

Minerals of high purity are usually formed in the Earth by a process that involves the flow of some liquid through cracks and pore spaces in the rock. In some particular circumstance of pressure, temperature or chemical surroundings, a component of such a fluid precipitates, and thus builds up a concentrated deposit in the pore spaces. Crystallization of the deposited substance may then further increase the purity, because the crystallization process may selectively build up the crystal from one set of atoms.

We certainly could not suppose that in an average mix of mantle rocks, a diamond could spontaneously form and grow to centimetre-size, from carbon atoms dispersed at a low concentration in the rock. Here, as in other mineralization processes, we have to suppose that a fluid containing carbon percolated through rock spaces and precipitated concentrated carbon. Veins of diamond would then be built up in these pore spaces, and a later eruption might bring fragments to the surface. What are the fluids that could be responsible for precipitating and concentrating the carbon?

The mere existence of the diamonds at these depths proves that unoxidized carbon exists there. The two types of fluids that one may consider for the concentration process would be carbon dioxide and methane, the latter possibly associated with heavier hydrocarbon molecules also.

Careful analysis of natural diamonds has provided an important piece of information. Tiny pore spaces in them have been analysed and found to contain small amounts of highly compressed gases, among which the carbon-containing ones were both carbon dioxide and methane (Melton and Giardini, 1974). It is clear, therefore, that not only unoxidized carbon, namely the diamond itself, but also methane, can exist down there. If both carbon dioxide and methane are present, and we are asking which one could shed and thereby lay down concentrated carbon, it would certainly have to be the methane. Carbon dioxide is much more stable against thermal dissociation than methane, and in any temperature regime where methane can partly survive, the carbon dioxide would not be dissociated at all; or in a regime where carbon dioxide would dissociate, there would not be any methane left. This observation strongly suggests, therefore, that methane is the fluid responsible for producing the concentrations of very pure carbon, which at these temperatures and pressures would crystallize to diamond. Heavier hydrocarbon molecules mixed with the methane may also be responsible.

Another important point is immediately settled by the observation of the existence of diamonds. It is that pore spaces in which fluids can flow exist at these depths, and that mineralization processes leading to great concentrations of certain substances can be active there, just as at shallower levels. Fluid pressures equalling the rock pressures seem to be widespread, at least in the crust and outer mantle, and this is a matter of great significance, both for the chemical processes and for the methods of ascent of fluids to the surface.

Does the presence of the diamonds and the associated fluids have anything to do with the process that gave rise to the violent eruptions that made the kimberlite pipes? Or are the two unrelated? In the first case we might suppose that these gases that seem to be down there were the ones that gave rise to the eruption. In the other case, we might suppose that diamonds are widespread or universal at this depth, more or less like coal-seams are at shallow depths, and that an unrelated eruption mechanism merely gives us an occasional sample of what is found down there. At the moment we cannot decide between these possibilities. In any case, the existence of the kimberlite pipes shows that high concentrations of gas can build up, and have been building up, and these concentrations can explode a hole through 150 kilometres of overlying dense rock. Quite large bubbles of high-pressure gas must have been assembled to do this, and only an inhomogeneous mantle containing volatile-rich materials could be responsible.

Do the kimberlite pipes tell us anything else about hydrocarbons? A. I. Kravtsov and colleagues (1976, 1981) in the Soviet Union have investigated just this aspect and have found that in the East Siberian diamond province the pipes contain bitumens and other hydro-carbons clearly specific to these pipes and not similar to such materials in the surrounding sediments. They can distinguish these deep-source hydrocarbons in the pipes, on the basis both of detailed chemical composition and of carbon isotopic composition. Bore holes in the pipes also frequently produced large amounts of methane, and methane was generally the dominant gas present. They concluded: "Thus kimberlite of the Udachnaya pipe contains bitumens and a complex assemblage of gases that include hydrocarbons and contain carbon with a substantially different isotopic distribution than bitumen, oil and gas in rocks of the sedimentary cover. Moreover, there are specific features of zoning in the distribution of bitumen and free gases, which are confined to fracture zones."

We see that the evidence from the diamonds is very simple and clear. Unoxidized carbon can and does exist in the outer mantle. It can be brought up without becoming oxidized; it is associated with a variety of hydrocarbon molecules, both within inclusions in diamond and also in other materials brought up in the eruptions. Volatile-rich regions exist in the mantle, so that high pressure gas bubbles become assembled there that can force their way violently through all the overlying rocks. This clearly shows that the Earth has an unmixed, inhomogeneous mantle, and that there is a high concentration of carbonaceous material in many areas of the globe.

4
EARTHQUAKES

The origin of diamonds has given us one glimpse into deeper levels in the Earth and it has shown us that there are fluids — liquids or gases — which flow there in pore spaces, where they distribute different minerals and can build up pressures large enough to erupt to the surface. The large amount of evidence associated with earthquakes can give us another type of information about occurrences at deep levels and, as we shall see, it also seems to point to the flow of fluids.

Whatever the details of any one earthquake may be, it is clear that earthquakes reflect some conditions at deep levels, showing that the Earth has not settled down to a completely quiescent state. Something is going on down there to produce the forces that move rock masses around at shallower levels, that perhaps cause rock masses to break and slide and shatter. The driving forces for all this are skill unknown in detail and there is a considerable amount of evidence that the motion of gases coming from deep levels has a lot to do with it.

The study of the phenomena associated with earthquakes makes clear that many of them could be considered as major eruptions, not like the eruptions of volcanoes or of the diamond pipes, concentrated in single locations, but like eruptions of gas, sometimes distributed over wide areas. Let us look in more detail at the wealth of evidence of earthquakes and the phenomena associated with them.

In olden times it used to be thought quite generally that earthquakes were caused by "vapours" that were confined deep in the Earth under great pressure and that occasionally burst through to the surface. Most of the classical literature concerning earthquakes is on these lines, from Aristotle onwards, and many detailed accounts of earthquakes in antiquity describe the emission of gases, as observed through flaming, smelling, bubbling in water or exploding at or around the time of the earthquake. We shall return to a description of this literature in more detail.

The present viewpoint of the nature of earthquakes seems to be quite different. In fact it was only around the year 1900 that the viewpoint changed and the earthquake literature adopted quite a different orientation. It was about then also that the seismograph was invented, and it allowed detailed deductions and calculations to be made, showing the movements of the ground that took place at an earthquake. This made it quite clear that the movements were those of a fracture of a stressed rock, breaking and suddenly sliding, and evidently discharging by this process some of the energy that had come to be stored as elastic strain energy in the rock. What it is that moves the crustal rocks around and builds up strain in them was not known, and it still is not known today. But in any case there is no doubt that strain is frequently built up in the rocks and that it is discharged sometimes in violent earthquakes.

The invention of the seismograph meant that earthquakes could be investigated in fine detail from the seismic records obtained, even in locations quite far away from the event. The subject of seismology flourished and produced a large amount of information, not only concerning the earthquakes themselves, but about the structure of the whole interior of the Earth.

The success of this line of investigation seems to have diverted attention from another aspect of earthquakes. We no longer find the eyewitness reports from the scene of the quake in the scientific literature, though we do find these sometimes in the popular press of the day. When we do have the eyewitness reports, they are very similar to all those gathered and recorded in the past. Eruptions, flames, noises, smells, asphyxiation, fountains of water and mud – all these are recurrent themes, as they were in antiquity. The earthquakes surely did not change their character. Only the investigators have shifted their attention.

Did the elastic rebound theory, as the new theory was called, really make it unnecessary to study these other phenomena? Were they merely small secondary effects, whereby fluids from the ground were expelled through the fissures that the earthquake opened up? Was the building-up of strain in the rock, leading up to eventual fracture, the underlying cause, and were all other phenomena only the consequence of this?

At best this would be a partial explanation only, so long as the causes of the build-up of strain or of the movement of crustal plates is not known.

While many of the major earthquake-prone regions of the Earth lie on prominent faultlines, along which some shear is noticeable, and for which the building-up of strain and a periodic slippage seems a good explanation, there exists also another type of seismic phenomenon for which such an interpretation seems inapplicable. This is the phenomenon of "earthquake spots".

In some areas that are not noted for general seismic activity, one may find a small spot of quite intense seismicity. It is often just a spot of a few kilometres, in which small earthquakes occur with very high frequency. One such spot, in the mountains of Northern Norway, was active for several years, and it was reported that an earthquake of an intensity that one could feel, occurred approximately once in any period of three days, while hundreds of microquakes could be registered in that time. The spot was only about 10 kilometres across, and the surrounding terrain showed no particular signs of seismic activity or shear displacement.

Similar spots have been noted in many other locations; there is one in Montana (near Flathead Lake), another in Arkansas, near the small town of Enola. Both of those have been active in recent times, and the one in Arkansas was also active some eighty years ago.

Another earthquake spot is on the north shore of the St Lawrence River, most interestingly just in the large meteorite impact structure ("astrobleme") called "Charlevoix". The large meteorite struck there some 350 million years ago, and detailed evidence of this impact has been obtained. Despite the length of time that has elapsed since then, it seems that even now the area has not settled down, and some activity is clearly still centred there. Earthquakes that can be felt occur every few years and microquakes are registered extremely frequently. In this case the proximity to the major fault line of the St Lawrence River complicates the discussion somewhat but, nevertheless, the concentration of the seismic activity to the 30 mile diameter impact area is quite evident.

Such spots clearly need a different explanation from that of shearing plates. Possibly their explanation has to do with gases forcing their way up and thereby causing fractures in the rock to open and shut repeatedly. In this respect these spots may be similar to another type of earthquake spot, namely an active volcano: there also, minor earthquakes are common, and they can clearly be attributed to the discharge of gases through magma channels. Where there are no magma channels, but only pathways through cool rock, the periodic ascent of gases may do nothing other than force some gaps apart and let them shut, causing the minor earthquakes.

We have investigated in some detail the Arkansas and the Charlevoix spots and in the course of this discovered that they both contain a most intriguing feature which sheds further light on this type of occurrence. This is the presence of clusters of earth mounds that stand abruptly out of the alluvial plain, from a few feet to 40 feet in height and up to 200 feet or so in horizontal dimensions. They are composed internally just of the clay and sand of the local alluvium and no good reason has been offered to account for their origin. In the Charlevoix region there are three clusters of such mounds, all in the alluvial fill of the ring depression that outlines the impact feature. In each case they lie on top of known fault lines and in one instance (at La Malbaie on the golf course) one of the clusters is an elongated feature precisely overlying the line of a fault.

The association of these strange mounds with locally concentrated seismic activity cannot reasonably be ascribed to chance. While such mounds do occur elsewhere, dense clusters of them are extremely rare and an explanation is needed. One cannot argue that the shaking of the ground of the earthquakes would itself cause this substantial extrusion from below, firstly because one would not understand how the volume of the subsurface would have been increased by that, and secondly because there are plenty of examples in other regions of much stronger earthquakes not having built up such features. The explanation must be similar to that of mud volcanoes, where the frequent and sometimes violent ascent of gases from below drives up the soft and sometimes fluid mud, and causes it to spill over the surface. As happens in the case of mud volcanoes, one of the mounds in the Charlevoix area has a funnel-shaped hole at the top where presumably gas and some sand were ejected.

Many of the known fields of large mud volcanoes are also "earthquake spots", with earthquakes sometimes related to massive gas eruptions. It seems reasonable therefore to regard these mounds as similiar phenomena but on a smaller scale and in a region of different subsurface conditions.

There are many other reasons for thinking that an explanation of all earthquakes in terms of the build-up of strain, the sudden fracture and the elastic rebound is not the whole story, even for the quakes that occur on lines of shear and for which there is little doubt that the main release of energy is due to the sudden discharge of built-up strain. One such line of evidence concerns the variety of precursory phenomena that occur characteristically months or weeks or days before many major earthquakes, and that cannot be accounted for adequately by the gradual build-up of strain. Another line of evidence comes from the large changes of volume of the ground that are sometimes involved, most often in the sense that the ground suffers a large collapse at the time of the quake. One must deduce in such cases that underneath there were pore spaces whose volume was drastically reduced in the quake. The large ocean waves – tidal waves or "tsunamis", which major earthquakes near shorelines have often caused – require for their explanation surprisingly large and sudden changes of volume in big areas of the crust of the Earth. No compression of solid rock or movement of liquid magma from below could account for such sudden large changes. A much more compressible substance has to be involved, and gas is really the only possibility.

A third line of evidence comes from the chemical nature of the gases that are expelled, which sometimes are quite out of keeping with what would have been expected as pore-fluids in the rocks of the region. When flames occur at the site of an earthquake, one might think, following the conventional viewpoint, that the flammable gases have come from hydrocarbon deposits in sedimentary reservoirs. But when flames are seen in areas of igneous rock, coming directly out of cracks, it is hard to accept such an explanation. Moreover, when we remember that in earthquake-prone regions tens of thousands of major earthquakes would have happened since the local rock was laid down, it is hard, according to the orthodox explanation, to believe that enough gas could have been incorporated to give similar phenomena every time, and that it would not have escaped in all the quakes of the past.

The earthquake literature in classical times

Earthquakes in Italy and in Greece are fairly common, and therefore there are many references to them in the classical literature of Greece and Rome. Volcanoes and earthquakes were the only sources of information about the deeper ground of the Earth. What was down there was clearly rather terrifying, and for this reason alone these phenomena attracted a lot of attention.

Aristotle, who dominated the scientific literature for 1800 years, sometimes with correct and sometimes with incorrect theories, provides the first detailed discussion of the earthquake process. According to him the theory that gases ("air") were responsible for earthquakes was first proposed by Anaxagoras, who said that "the air, whose natural motion is upwards, causes earthquakes when it is trapped in hollows beneath the Earth . . . ". In his review of the earthquake literature of the time, Seneca, around A.D. 63, stresses that "It is a favourite theory of most of the greatest authorities that it is moving air which causes earthquakes".

Why did the ancients favour air, or gas as we would call it today, as the active element in earthquakes? For one thing, they believed that there was a close connection between seismic and volcanic phenomena, and that volcanic eruptions, which clearly involve gas, acted as a safety valve for the forces that generated earthquakes. The presence of subterranean air was generally accepted in antiquity. Seneca had "No doubt that a great quantity of air lies within the Earth and a widespread atmosphere occupies the hidden spaces underground". Despite much confusion about the action of underground gases, there appears to have been some observational basis for the theory. It had been noted that there were warnings on occasions before violent earthquakes that "in winter it turns too hot, and in summer along with a tendency to haze, the orb of the Sun presents an unusual colour. . . . Springs of water generally dry up; great flames dart across the sky . . . furthermore there is a violent rumbling of winds beneath the Earth . . . " (Pausanias in his description of Achaia).

Seneca also noted that "Before the earthquake a roaring noise is usually heard from winds that are creating a disturbance underground". Pliny, in his *Natural History*, also discusses the earthquake phenomenon and the precursory effects, and he notes that a sign of impending earthquakes is that "water in wells is muddier and has a somewhat foul smell, and that there is a remedy for earthquakes, such as is frequently afforded by caves, as they supply an outlet for the confined breath. This is noticed in whole towns. Buildings pierced by frequent conduits for drainage are less shaken and also among these the ones erected over vaults are much safer, as is noticed in Italy at Naples. . . ." Seneca observed that "often when an earthquake occurs, if only some part of the Earth is broken open, a wind blows from there for several days, as happened – according to reports – in the earthquake which Chalcis suffered. . . . Clearly the wind had made for itself the passageway through which it moved."

Seneca, quoting from the lost work of Callisthenes, related that "among the many prodigies by which the destruction of the two cities Helike and Bura was foretold, especially notable were both the immense columns of fire and the Delos earthquake". Unfortunately, the time and place of the columns of fire connected with the earthquake are unclear. There does exist, however, an account by Aelian of the general exodus of ground-dwelling animals from Helike, said to have begun five days before that same earthquake.

Seneca was moved to write his work on earthquakes by a seismic shock that wrecked Pompeii some 16 years prior to the even greater disaster of the eruption of Vesuvius. And he did provide one peculiar detail from that event to support the gas eruption theory:

> I have said that a flock of hundreds of sheep was killed in the Pompeian district. . . . The very atmosphere there, which is stagnant . . . , is harmful to those breathing it. Or, when it has been tainted by the poison of the internal fires and is sent out from its long stay it stains and pollutes this pure, clear atmosphere and offers new types of disease to those who breathe the unfamiliar air. . . . I am not surprised that sheep have been infected – sheep which have a delicate constitution – the closer they carried their heads to the ground, since they received the afflatus of the tainted air near the ground itself. If the air had come out in greater quantity it would have harmed people too; but the abundance of pure air extinguished it before it rose high enough to be breathed by people.

Then, in connection with the same earthquake, Seneca sought to explain the series of aftershocks, felt for several days throughout Campania, in terms of the gas eruption theory. He concluded that not all the air had yet left, but was still wandering around, even though the greater part had been emitted. He thought that after the largest shock there would only be gentle quakes, because the first tremor had created an exit for the struggling air: "The remains of the air do not have the same force, nor do they have any need to struggle, since already they have found a path and follow the route by which the first and largest force of air escaped".

We would not discuss these historical records were it not for the fact that eyewitness descriptions since then, and from many parts of the world, repeat the same themes. Rumblings and hissing noises during earthquakes, sulphurous fumes, changes in ground water levels, hot gases and flames recur in many of the descriptions. Sir Isaac Newton (1730) also subscribed to the view that earthquakes were connected with gases. He wrote that "sulphureous Steams abound in the Bowels of the Earth and ferment with Minerals, and sometimes take fire with a sudden Coruscation and Explosion, and if pent up in subterraneous Caverns, burst the Caverns with a great shaking of the Earth, as in springing of a Mine". Furthermore, the first edition of the *Encyclopaedia Britannica* in 1771 contained the entry: "EARTHQUAKE, in natural history, a violent agitation or trembling of some considerable part of the earth, generally attended with a terrible noise like thunder, and sometimes with an eruption of fire, water, wind, &c. See PNEUMATICS".

It was John Michell (1761), a brilliant scientist of the eighteenth century, who contributed two major items to the understanding of earthquakes. He was the first to recognize that the vibratory motion in earthquakes was due to the propagation of elastic waves in the Earth's crust, similiar to sound waves in air, but with a higher velocity. He was, however, unaware – and it was not generally accepted until the twentieth century – that these waves are generated by the fracture of stressed rock.

Michell also recognized another kind of earthquake disturbance, consisting of slow, ocean-like waves that could actually be observed moving along the surface of the ground. These "visible waves" cannot be explained adequately in terms of elastic wave motion, and there is not much discussion of them in modern seismological texts. Michell attempted to explain them in terms of an eruption of vapour, and that may indeed be the best explanation. What would happen if a burst of high-pressure gas from a depth of many kilometres, and therefore with a pressure of thousands of atmospheres, were suddenly released through fissures in the bedrock into a region beneath a relatively impervious layer of 'soil, which is not brittle enough to develop fissures? Michell reasoned as follows:

> Suppose a large cloth, or carpet, (spread upon a floor) to be
> raised at one edge, and then suddenly brought down again to
> the floor, the air under it, being by this means propelled, will
> pass along, till it escapes at the opposite side, raising the cloth
> in a wave all the way as it goes. In like manner, a large quantity
> of vapour may be conceived to raise the earth in a wave, as it
> passes along between the strata, which it may easily separate in
> an horizontal direction, there being . . . little or no cohesion
> between one stratum and another.

The evidence for the phenomenon of visible waves in numerous earthquakes in ancient and modern times is quite indisputable. Where an earthquake is felt both on exposed basement rock and on alluvial fill, the visible waves are reported on the alluvium only. In many cases the large displacements of these waves seem to have created more destruction than the sharp shocks of the quake. It is likely that the soil of the area is genuinely lifted off the basement rock and that it is therefore subject to flexural gravity waves, just like the carpet in Michell's example.

Michell's attention was directed to earthquakes as a result of the disastrous one that struck Lisbon in 1755, and he drew from a large number of eye-witness accounts that appear to link these earthquakes with gas. He writes about the flames from the Earth and the peculiar fog accompanying the Lisbon earthquake, and he also describes precursors of earthquakes in Jamaica and New England that occurred two or three days before and involved waters of wells being rendered muddy and developing a sulphurous smell. For the earlier Jamaican earthquake (of 1688) he records that the ground was seen to rise in waves like the sea; and that these waves could be distinguished to distances of some miles by the motion of the tops of the trees. As we shall see, this is a common phenomenon during major earthquakes, and on at least one occasion has been clearly recorded on film.

Other investigators followed Mitchell's explanations and extended them further. In America Professor Samuel Williams (1785), in a now largely forgotten paper that deserves quotation, writes as follows:

> What thus moved under and hove up the surface of the earth, was probably a strong elastic vapour. This is inferred from the phenomena that . . . preceded the earthquakes, and looked like a previous preparation. In the [New England] earthquakes of 1727 and 1755, in particular, it was evident, that the causes by which they were produced, were at work several days before they became ripe for an explosion. As tho' some grand fermentation was taking place in the bowels of the earth, the water, in several wells and springs, was uncommonly altered in its motion, colour, smell and quality. . . . Nothing could better agree with the origin and production of a subterraneous elastic vapour, than this circumstance. For however such a vapour be generated, by mixture, fermentation or fire, it would require some previous preparation, for its production, or before it would be collected in sufficient quantities to cause an explosion, or acquire sufficient force to move and shake the surface of the Earth.
>
> The noise or roar, occasioned by the earthquakes, has always been such as might have been expected from a subterraneous

vapour, when fiercely driving along under the surface of the
earth. What report might be expected from a strong elastic
vapour, when its motion is confined and directed by a particular
channel or passage, we may learn from that of a blazing
chimney . . . and there is nothing to which the report of our
earthquakes is more similiar, or has been more often compared.

There have been other effects upon the water, such as a
surprising flux and reflux of the sea, extraordinary agitations
and commotions of the waters, – an uncommon destruction of
fish, &c . . . And they seem to be plain and evident marks and
effects, of the discharge of the subterraneous vapours, at the
bottom of the sea. Such a discharge, when small, would be
sufficient to occasion the destruction of such fish as were near
it: and when large, would put an end to the earthquake, and
produce the most extraordinary agitations and commotions of
the sea, by a furious eruption of vapours at its bottom; which
would immediately force their way through, or carry up before
them, the whole body of water that lay over them.

And thus as to the conclusion: – it might be naturally
expected, that as the vapours, by which the earthquakes were
caused, were some time in growing ripe, fermenting, or in a state
of previous preparation, they would not be wholly spent at the
first shock: but what has remained and what has gathered after
a great explosion, has produced various small shocks in several
places, for some time after the great ones: – thus wasting and
evaporating by little and little, as they were collected and
prepared at first; till, by degrees, all has become quiet again.

Clearly, Williams was also impressed by the observations, and
was not merely repeating an ancient revived opinion regarding
earthquakes.

And well into the next century, many scientific works dealing with
the cause of earthquakes continued to mention the evidence for the
associated gas eruption phenomena. For example, Bischof (1839)
wrote that

earthquakes may also be produced by gaseous exhalations in the
interior of the globe. At least in many accounts of earthquakes,
mention is made of the exhalation of gases from rents produced
by them, and the smell of sulphuric acid, and of sulphurous
vapours, which indicate the presence of sulphuretted hydrogen
[hydrogen sulphide]. These last may have occasioned also the
destruction of the fish in the sea, and in lakes, during
earthquakes; many instances of which are known. The bursting
forth of flames from the earth and from the sea, which is so

often mentioned, also indicates the presence of inflammable gases . . . . The heating and boiling up of the water in the sea and in lakes, the spouting up of streams of water, as well as the ejection of various substances from fissures in the earth, which have occasionally been witnessed, may be satisfactorily explained by the rising of steam and gases, which may have the effect either of heating the water, or of throwing out solid bodies.

The British seismologist Robert Mallet (1852, 1853, 1854) published a 600-page catalogue, of more than 5000 historical earthquakes, with descriptions of accompanying phenomena, including luminous phenomena (earthquake lights), explosions, peculiar atmospheric conditions, and so on. In his textbook on earthquakes, John Milne (1886), the inventor of the modern seismograph, concluded that most earthquakes were caused by the explosive effect of steam in a process related to volcanism. And the great naturalist Alexander von Humboldt (1822) summed up a widely accepted nineteenth-century view on the subject: "Everything in earthquakes seems to indicate the action of elastic fluids seeking an outlet to diffuse themselves in the atmosphere".

The Italian investigator Galli (1911) collected accounts of 148 cases of major earthquakes, during which luminous phenomena were seen, many involving flames issuing from the ground. In modern Japan many photographs have been taken of earthquake lights (it is a country with a high incidence of earthquakes and a large amount of photographic equipment), and while there may be other luminous effects, probably also caused by the emission of gas, flames hovering over the ground seen from various distances seem to be the main explanation for the lights.

The descriptions available from many historical earthquakes are so startling that it is worthwhile reproducing some of them here. The descriptions repeat so consistently a similar set of phenomena that one has to take them seriously. In most cases invention and collusion cannot be suspected, since many of the reporters would not have had access to reports of earlier occurrences. These descriptions make major earthquakes look much like violent eruptions, perhaps similar to gas eruptions from volcanoes or mud volcanoes, except that they occurred in areas where neither lava nor mud was available to be expelled. The airborne noises, the flames, the air pollution are all similar, and while most of the intense effects take place at the time of the quake, some of the effects occur as precursors and cannot therefore be ascribed to secondary effects of the mechanical deformation of the ground.

To see something of the consistency of the descriptions from numerous major earthquakes, it is worth looking at a selection, and we give here brief excerpts in chronological order from some 18 major earthquakes in historical and modern times, selected from among hundreds that are on record.

QUEBEC, 5 FEBRUARY 1663

"Pikes and lances of fire were seen, waving in the air, and burning brands darting down on our houses – without, however, doing further damage than to spread alarm wherever they were seen." (Lalement, 1663)

"From the [earth] emanated fiery torches and globes of flame – now relapsing into the earth, now vanishing in the very air, like bubbles . . . a great section of the earth [was] borne upward and carried into the [St Lawrence] river; and, at the place whence it was separated by the yawning open of the earth, there burst forth globes of smoke and flame, at certain spaces from one another, and very dense clouds of ill-smelling ashes were cast upward; and as these fell down, the deck of [a] ship was filled with them." (Simon, 1663)

VOSGES (FRANCE), 13 MAY 1682

"Flames were also seen coming from the earth, without there appearing any opening, nor any other outlet, except in a single spot, where there opened a cleft, the depth of which could not be measured. . . . The flames which issued from the earth, and which occurred most frequently in places that had been planted, such as woods, in no way burned the objects they encountered; they gave off an extremely disagreeable odor, but one that had nothing sulfurous about it." (Quoted by Galli, 1911)

NORCIA AND AQUILA (ITALY), 14 JANUARY AND 2 FEBRUARY 1703

"In Aquila and Norcia, and in other places . . . the earth was here and there observed to split in cracks, from which streamed the evil odors of sulfur and bitumen; and men in Aquila most worthy of trust write that in many places after the earthquake sulfur and fire issued from the opened earth." (Quoted by Galli, 1911)

LISBON, 1 NOVEMBER 1755

". . . we began to hear a rumbling noise, like that of carriages, which increased to such a degree as to equal the noise of the loudest cannon; and immediately we felt the first shock, which was succeeded by a second and a third; on which, as on the fourth, I saw several light flames of fire issuing from the sides of the mountains, resembling that which may be observed on the kindling of coal. . . . I observed from one of the hills called the Fojo, near the beach of Adraga [near Colares], that there issued a great quantity of smoke, very thick, but not very black; which still increased with the fourth shock, and after continued to issue in a greater or less degree. Just as we heard the subterraneous rumblings, we observed it would burst forth at the Fojo; for the quantity of smoke was always proportional to the subterraneous noise." (Stoqueler, 1756)

KOMAROM (HUNGARY), 28 JUNE 1763

"Ruptures in the soil originated in thousands of places. From almost all of them water and quicksand were emitted together with flames and stinking smoke. . . . The river Danube began to rise . . . and the water appeared to be steaming as though boiling. It had a sulphurous smell. The majority of the ruptures occurred near the river bank and from some of them flames emerged alternately with the sand and smoke. Fertö Lake, 100 km west of Komarom, began to rumble and foam very intensely. . . . Flames as big as a barrel were seen over the river itself. Many horned cattle perished in the terrible stinking vapour that came from the earth. . . . At the bank of another smaller river, the Vag, red-colored flames rushed up from the ruptures, followed by sulphurous waters. . . . At some places the waters that came from the earth were distinctly black. The water of the river Vag appeared to be boiling." (Quoted by Rethly, 1952)

CALABRIA, 5 FEBRUARY 1783

"At the time of the earthquake, during the night, flames were seen to issue from the ground in the neighborhood to the city towards the sea, where the explosion extended, so that many countrymen ran away for fear; these flames issued exactly from the place where some days before an extraordinary heat had been perceived." (Ippolito, 1783)

"The water of the wells, of the sea, and also of the fishponds, a few hours before the earthquake of 5 February struck in Cosenza and neighboring villages, was seen to raise its level, all foaming as though boiling, without being observed to have a greater heat than normal." (Quoted by Galli, 1911)

CUMANA (VENEZUELA), 14 DECEMBER 1797

"At Cumana, half an hour before the catastrophe . . . a strong smell of sulfur was perceived near the hill of the convent of St.

Francis. . . . At the same time flames appeared on the banks of the Manzanares, near the hospice of the Capuchins, and in the gulf of Cariaco, near Mariguitar." (Humboldt, 1822)

NEW MADRID, 16 DECEMBER 1811, 23 JANUARY 1812, AND 7 FEBRUARY 1812

"During the first four shocks, tremendous and uninterrupted explosions, resembling a discharge of artillery, were heard from the opposite shore [of the Mississippi River]. . . . Wherever the veins [fissures] of the earthquake ran, there was a volcanic discharge of combustible matter to a great height, an incessant rumbling was heard below, and the bed of the river was excessively agitated, whilst the water assumed a turbid and boiling appearance. Near our boat a spout of confined air, breaking its way through the waters, burst forth, and with a loud report discharged mud, sticks, etc., from the river's bed, at least 30 feet above the surface." (Pierce, 1812)

"In some places there issued from the earth something like wind from the tube of a bellows passing through burning coal. . . . In some places west of the Mississippi a troublesome warmth of the earth was perceptible to the naked feet. The next day but one before the first earthquake was darkened from morning to night by thick fog and divers persons perceived a sulphureous scent. . . . Explosions like the discharge of a cannon at a few miles' distance were heard; and at night, flashes of lightning seemed sometimes to break from the earth. . . . The ponds of water, where there was no wind, had a troubled surface, the whole day preceding any great shock. . . . A dull and heavy obscuration of the atmosphere usually preceded the shocks. The effluvia which caused the dimness of the day seemed to be neither cloud nor smoke, yet resembling both." (Haywood, 1823)

". . . A loud roaring and hissing . . . like the escape of steam from a boiler, accompanied by . . . tremendous boiling up of the waters of the Mississippi in huge swells . . . flashes such as would result from an explosion of gas . . . complete saturation of the atmosphere with sulphurous vapor . . . the earth was observed to roll in waves a few feet high with visible depressions between . . . these burst, throwing up large volumes of water, sand, and coal." (Fuller, 1912)

LIMA, 30 MARCH 1828

Water in the bay "hissed as if hot iron was immersed in it," bubbles and dead fish rose to the surface, and the anchor chain of *HMS Volage* was partially fused while lying in the mud on the bottom. (Bagnold, 1829)

VALDIVIA, 20 FEBRUARY 1835

". . . the bay of Concepcion was agitated by great waves . . .
Two explosions, or eruptions, were seen while the waves were
coming in: one beyond the island of Quiriquina . . . appeared to
be a dark column of smoke, in shape like a tower. Another
rose in the middle of the bay of San Vicente, like, the blowing
of an immense imaginary whale. . . . At the time of the ruin,
and until after the great wave, the water in the bay appeared to
be everywhere boiling; bubbles of air, or gas, were rapidly
escaping. The water also became black, and exhaled a most
disagreeable sulphureous smell. . . . The island of Juan
Fernandez was affected very much. Near Bacalao head, an
eruption burst through the sea in a place about a mile from
the land, where the depth is from fifty to eighty fathoms. Smoke
and water were thrown out during the greater part of the day:
flames were seen at night." (Fitz-Roy, 1836)

FT. YUMA (COLORADO DELTA), 29 NOVEMBER 1852

" . . . the earth, after moving backwards and forwards some
three or four feet in an undulating way, nearly capsizing men,
mules and wagons, burst with a loud report, resembling much
a heavy peal of thunder, rending the earth and leaving a long
and deep chasm, from which exuded a large volume of gaseous
matter resembling the smoke of an overheated furnace." (San
Francisco *Daily Alta California*, 31 December 1852; quoted by
Agnew, 1978)

ARICA (CHILE), 13 AUGUST 1868

"From every fissure there belched forth dry earth like dust, which
was followed by a stifling gas . . . which severely oppressed
every living creature, and would have suffocated all these if it
had lingered longer stationary than it did, which was only
about 90 seconds." (New York *Tribune*, 14 September 1868)

OWENS VALLEY (CALIFORNIA), 26 MARCH 1872

"People living near Independence . . . said [that] at every
succeeding shock they could plainly see in a hundred places at
once, bursting forth from the rifted rocks great sheets of flames
apparently thirty or forty feet in length, and which would coil
and lap about a moment and then disappear." (San Francisco
Chronicle, 2 April 1872)

"Immediately following the great shock, men whose judgment
and veracity is beyond question, while sitting on the ground
near the Eclipse mine, saw sheets of flame on the rocky sides of
the Inyo mountains but a half a mile distant. These flames,

observed in several places, waved to and fro apparently clear of
the ground, like vast torches; they continued for only a few
minutes." (Inyo *Independent*, 20 April 1872)

SONORA (MEXICO), 3 MAY 1887

"Another effect of the earthquake which terrified the frightened
inhabitants of these places, was the fire upon all the mountains
around the epicenter and even some situated in the territory of
Arizona, among others the ridge of San Jose. Some of these it
is said continued in flames for many days." (Aquilera, 1920)

"The Sierra Madre fires were, beyond question, synchronous,
and arose similarly. The evidences of gaseous irruption were few
but striking. Primarily were the statements of many who claim
to have seen streaks of flame at different points. . . . [Further
evidence] consisted of cinders about the margins and on the
walls of the river-fissures, and the discovery of burnt branches
overhanging the edges of such places, as well as the same
testimony on some of the hills and mountains near the main
fault." (Goodfellow, 1888)

SWABIA (SOUTHERN GERMANY), 16 NOVEMBER 1911

The following are from among the many eyewitness accounts quoted
by Schmidt and Mack (1913):

"We saw a sea of flames, gas-like and not electrical in nature,
shoot up out of the paved market street. The height of the
flames I can estimate at 8 to 12 cm; it was like when you pour
petroleum or alcohol on the ground and light it."

"I observed very precisely how a bright fire, which had a
bluish color, came out of the ground in the meadow. Its height
was about 80 cm. . . . The fire was present not only in the
meadow but also in the whole surroundings of our house."

"Some people in the streets . . . noticed that for a while before
the quake and particularly after it an evil stuffy air made
breathing almost impossible."

"We were very well the whole evening when, shortly before
the first tremor, perhaps five minutes before, a giddiness took
hold of me, it felt as if the surroundings were receding and I
was suddenly unwell to such an extent that I wanted to go to
bed, thereby noticing that I was unsteady on my legs. At that
moment the earthquake happened and afterwards I was again
completely well, despite the scare."

RUMANIA, 10 NOVEMBER 1940

The following are phrases used in eyewitness accounts collected by
Demetrescu and Petrescu (1941):

" . . . a thick layer like a translucid gas above the surface of
the soil . . . irregular gas fires . . . flames in rhythm with the
movements of the soil . . . flashes like lightning from the floor
to the summit of Mt Tampa . . . flames issuing from rocks, which
crumbled, with flashes also issuing from non-wooded
mountainsides."

HAICHENG (CHINA), 4 FEBRUARY 1975
In the weeks prior to this earthquake in northeast China, the air
temperature in the vicinity of the fault was higher than in the
surrounding region, and this difference increased at an
accelerated rate up to the day before the earthquake, when it
reached 10°C.

"During the month before the quake a gas with an
extraordinary smell appeared in the areas including Tantung
and Liao-yang. This was termed 'earth gas' by the people . . .
one person fainted because of this. . . . Many areas were
covered with a peculiar fog (termed 'earth gas fog' by the people)
just prior to the quake. The height of the fog was only 2 to 3
meters. It was very dense, of white and black color, non-uniform,
stratified and also had a peculiar smell. It started to appear 1
to 2 hours before the quake and it was so dense that the stars
were obscured by it. It dissipated rapidly after the quake. The
area where this 'earth gas fog' appeared was related to the fault
area responsible for the earthquake." (Liao-ling Province
Meteorological Station, 1977)

We advance here a tentative explanation. If this earthquake
was triggered by the ascent of deep-source gases which
embrittled the stressed rock, then these same gases may
previously have mixed with, and driven out of the overlying
soil, some of the gases normally present in the porosity above
the water table. The soil gases below a few meters' depth will
have been about 10°C warmer than the surface air in midwinter
and saturated with water vapor. They would therefore produce
a fog on contact with the surface air. Despite its warmth, the
gas mixture would remain close to the ground if it were more
than about 7% CO_2. Such a CO_2 content of air is sufficient to
cause people to feel unwell. An emission of gases may also
have accounted for the several incidents of anomalous animal
behavior and changes in groundwater reported prior to this
earthquake (Deng *et al.*, 1981).

SUNGPAN-PINGWU (CHINA), 16, 22, AND 23 AUGUST 1976
"From March of 1976, various macroscopic anomalies were
observed over a broad region. . . . At the Wanchia commune
of Chungching County, outbursts of natural gas from rock

fissures ignited and were difficult to extinguish even by dumping dirt over the fissures. . . . Chu Chieh Cho, of the Provincial Seismological Bureau, related personally seeing a fireball 75 km from the epicenter on the night of 21 July while in the company of three professional seismologists. . . . At another place, a fireball started near a house, rose up along an arbor, and burned a hole in the roof of the house. A total of about 1000 fireballs were sighted, 50 in one evening. During the daytime, small smoke balls were reported, presumably representing the same phenomenon as the fireballs seen at night. . . . More fireballs occurred along intersections of river beds and fractures. . . . A few hours before the Sungpan-Pingwu earthquakes, a technical member of the provincial seismological [station] was surprised to find that the water in the well at the station was bubbling violently. . . . Some people experienced nausea before the Sungpan-Pingwu and Tangshan earthquakes." (Wallace and Teng, 1980)

The San Francisco earthquake

The earthquake that destroyed parts of San Francisco and virtually all of Santa Rosa occurred at 5:12 a.m. on 18 April 1906. It was most intense perhaps a hundred kilometres north of San Francisco. We will here list some excerpts from the numerous reports, all indicating violent gas emission from the ground, gases that contained the poisonous hydrogen sulphide and gases that were frequently flammable. It is the earthquake for which the most detailed reports exist, and which shows every type of phenomenon which we have noted in other cases.

(a) EFFECTS IN AIR

An extensive list of noises heard at the time of the shock, compiled from witnesses by Lawson and others (1908), includes the following: From Santa Rosa, "Heard noises in SW; then felt breeze; then felt shock". From Cotati, "Sound as of a strong wind before the shock". From Point Reyes Station, "Heard roar, then felt wind on my face". From Calistoga, "A rushing noise before the shock came". From Pescadero, "Noise as of wind preceded the shock". And from Mount Hamilton, "Sound as of flight of birds simultaneously with shock".

Other clear evidence for gas is given by a report published on 23 April in the Santa Rosa *Democrat-Republican* (the first newspaper to appear after the devastation). It said:

J.B. Doda, who came over from Fort Ross on Monday, reports
that the earthquake caused immense cracks in the earth there,
from which strong gases are emitted which make men and cattle
sick.

Also, according to Edgar Larkin (1906), who collected a great many
accounts, the odour of hydrogen sulphide was noted in the area of
Sausalito. He also reported that

sulfurous odors were pungent in Napa County during the night
of the 17th and 18th before the upheaval, and lasted all
day. . . . From many of the letters it is clear that the entire region
north and east of San Francisco is saturated with gases of sulfur
origin. . . .

In Santa Rosa, according to Lawson and others (1908), a strong
smell of sulphur had been noticed two days before the earthquake
by one Charles Kobes. Since during an earthquake eight years
previously, "sulfur fumes came up from under his house which
almost drove his family from home", the recurrence of this phenomenon on 16 April 1906 caused Kobes to tell his family that there
would be another earthquake.

(b) EFFECTS IN WATER

Numerous indications of hydrogen sulphide in bodies of water were
reported. According to Larkin (1906), "creeks became milky in
several places as if gas escaped from the water". Hydrogen sulphide
bubbling through water is known to give it a milky appearance.
Another report in the San Jose Herald of 2 May 1906 states that in
Monterey Bay, on the day of the quake, there were thousands of
strange fish floating on the water a few miles offshore, none of which
were known to old fishermen on the boat. Similar reports of massive
fish kills at times of earthquakes, especially of bottom-dwelling fish,
are known from Japan, in some cases also associated with the
description of milkiness of the water. Again, hydrogen sulphide,
which is highly toxic to fish, seems a likely explanation, and in each
case it is bottom dwelling fish which are not normally caught that
are the chief victims.

(c) ANOMALOUS ANIMAL BEHAVIOUR

Strange animal behaviour preceding earthquakes is well documented
in many parts of the world. Dogs, pigs, horses, cows and many other

animals seem to show signs of restlessness or extreme disturbance prior to major earthquakes, and I would attribute this to their ability to smell the outflow of ground gases much more readily than humans and to be altogether much more concerned about smells. In San Francisco the major reports of this nature concerned the behaviour of dogs (Lawson *et al.*, 1908), which are reported to have been howling during the night preceding the earthquake.

(d) EARTHQUAKE LIGHTS

Again, as in many other earthquakes, there are many reports of flames issuing from the ground, either seen close-by or seen as a glow of light in the distance. In fact, while it was reported that the great fire, which was initiated by the earthquake, was in part caused by broken gas mains in the streets of San Francisco, this may not have been the major cause. There are numerous reports of flames seen in neighbouring areas where no gas mains existed. Thus, George Madeira, a veteran mining engineer from Healdsburg, reports in the Santa Rosa *Republican* for 4 April 1910:

> While investigating the natural phenomena of the seismic
> disturbance of April 18, 1906, I visited the mountain ranch of
> Mr and Mrs Adams, a mile and one-half northeast of Cazadero.
> They stated that for two nights preceding the earthquake they
> "had seen small streams of lightning running along the ground".
> Their attention was called to the phenomenon by the incessant
> barking of their dog.

Here, evidently some 30 hours before the shock, earthquake lights were reported in what was soon to be the epicentral region.

During the earthquake itself there were more such accounts, like that of J. E. Houser, an engineer in San Jose, California, quoted by Larkin (1906):

> On April 18, I awakened five minutes before our clock struck
> five. I heard a rumbling noise as of distant thunder. Two mares
> with young colts were running and whinnying in an adjacent
> lot, in alarm as though dogs were after them. Dogs were there,
> but they too gave unusual warning of danger. At 5:12 my bed
> jumped from under me, the movement starting from a
> standstill.
>   The force seemed to raise up the house and turn it to the right
> upward and left downward, with tremendous power, so

forcible as to tear me loose from the door frame to which I was
clinging with both hands, my wife holding around my waist.

We both could see down Alameda Street, looking eastward,
and we both saw the whole street ablaze with fire, it being of
a beautiful rainbow color, but faint. We passed out into the
street and met a man who asked, "Did you see the fire in
Alameda Street?" An hour later a friend told me that the ground
all around was a blaze of fire.

That this was not an isolated instance is clear from the further
accounts gathered by Larkin (1906). Unfortunately he is much less
precise here on details:

> . . . a letter from a point north of San Francisco describes blue
> lights as flickering like an Aurora, over a wide area of
> marshland, with a troubled surface of adjoining water. . . . In
> Petaluma Creek the water splashed up as though thousands of
> stones were dropped into it; and blue flames eighteen inches in
> height played over a wide expanse of marshland. . . . At 5 p.m.
> [*sic*, may mean 5 a.m.] before the turbulence, "A flickering
> luminous haze" was seen playing above the ground. . . . Blue
> flames were seen hovering over the bases of foothills in Western
> San Francisco.

(e) EXPLOSIVE NOISES (BRONTIDES)

According to George Madeira in a letter written on 5 May 1908,
as quoted by Alippi (1911),

> Explosions much resembling the discharge of heavy guns have
> for the past two years been heard at intervals in the West and
> Middle Coast range of mountains, particularly in Marin,
> Sonoma and Mendocino Counties. Heavy detonations and
> rumblings were heard near the base of Mt Tamalpais, Marin
> County, during the winter months and previous to the great
> earthquake which destroyed San Francisco and Santa Rosa in
> Sonoma County April 18th, 1906, and have been heard at
> stated times up to this writing.

Some of these later explosions evidently accompanied earthquake
aftershocks.

(f) VISIBLE WAVES

The phenomenon of slowly rolling waves, like the waves at sea, was
reported from many places in the San Francisco earthquake. Lawson
and others (1908) list over twenty such accounts distributed

geographically from the vicinity of Eureka to Visalia, a distance of more than 600 kilometres. Several of these accounts explicitly compare the ground motion observed to that of waves in the ocean.

We see that these descriptions make major earthquakes look much like violent eruptions, quite similar to gas eruptions from volcanoes or mud volcanoes. The airborne noises, the flames, the air pollution are all similar, and while most of the intense effects take place at the time of the quake, some of the effects occur as precursors and cannot therefore be ascribed to secondary effects of the mechanical deformation of the ground.

How do animals predict earthquakes?

Accounts of strange animal behaviour prior to earthquakes come from all parts of the world, from modern times as well as from classical antiquity. The earliest such description I could find contains the earthquake which totally destroyed the Greek cities of Helike and Bura on the southern coast of the Gulf of Corinth in the winter of 373-374 B.C. The Roman writer Aelian (c. A.D. 200) in his book *On the Characteristics of Animals* tells the following remarkable story:

> For five days before Helike disappeared, all the mice and martens and snakes and centipedes and beetles and every other creature of that kind in the town left in a body by the road that leads to Carynea. And the people of Helike, seeing this happening, were filled with amazement, but were unable to guess the reason. But after the aforesaid creatures had departed, an earthquake occurred in the night; the town collapsed; an immense wave poured over it, and Helike disappeared, while ten Lacedaemonian vessels which happened to be at anchor close by were destroyed together with the city I speak of.

Aelian's rather quaint description of an organized exodus of all the town's vermin is no doubt an exaggeration. He was, after all, writing nearly six centuries after the events described, more than enough time for facts to take on the embellishments of folklore. Nonetheless, I believe that this story was not simply made up. Rather, it seems very likely that *some* highly unusual disturbance of ground-dwelling creatures must have made an impression on the people of Helike prior to the earthquake and tsunami that destroyed their city and the Spartan ships. The reason for taking this story seriously is that we know of hundreds of accounts of animals behaving in a similar fashion prior to earthquakes. They are reported in circumstances as remote from one another in space and time as ancient Greece and modern China.

To cite a recent example, we have the following account from an eyewitness to the catastrophic Tangshan (China) earthquake of 28 July 1976. The author (Li, 1980) and his companions were all intellectuals in a "re-education program" at a state-owned farm outside Tangshan. The time was around midnight, some four hours before the earthquake:

> We were telling stories in the dormitory yard when out of the
> large dorm opposite ours burst hundreds of rats. Back and
> forth they swarmed, many scrambling 5 or 6 feet up the walls
> until they lost hold. All we could do was watch until they
> finally vanished into the darkness. As we pondered this in
> amazement, the sound of thousands of excited hens and
> roosters reached our ears. There was a poultry farm nearby, but
> nobody had recalled ever hearing the roosters crow at night.
> None of us knew that this queer animal behaviour foretold the
> coming of an earthquake . . .

Though filled with amazement – like the people of Helike twenty-three centuries before – they had no suspicion of impending disaster. So they went to bed, and a few hours later some of them were killed when their dormitory collapsed. More than 200,000 people died in the Tangshan earthquake. The subject of animal precursors is one that must be taken seriously.

A summary of the literature on unusual animal behaviour prior to earthquakes was prepared by Lee and others (1976). They believe that many of the individual reports, which usually lack the objectivity of scientific observation, are unreliable. In this they are certainly correct, but they also recognize that the similarities in the descriptions among widely separated cultures suggest an underlying basis in fact.

The reports of strange animal behaviour before earthquakes cover a wide spectrum of animal varieties (von Hentig, 1923). An extraordinary account of the Naples earthquake of 26 July 1805, by an anonymous Italian writer, describes accurately and concisely almost every one of the most frequently mentioned types of precursory

animal behaviour. The account, quoted by Wittich (1869), is as
follows:

> Some minutes before [the shocks] were felt, the oxen and cows
> began to bellow; the sheep and goats bleated, and, rushing in
> confusion one on the other, tried to break the wicker-work of
> the folds; the dogs howled terribly; the geese and fowls were
> alarmed and made much noise. The horses, where fastened in
> their stalls, were greatly agitated, leapt up, and tried to break
> the halter with which they were attached to the mangers; those
> which were proceeding on the roads suddenly stopped and
> snorted in a very strange way. The cats were frightened and
> tried to conceal themselves, or their hair bristled up wildly.
> Rabbits and moles were seen to leave their holes; birds rose as
> if scared from the places on which they had alighted; and fish
> left the bottom of the sea and approached the shores, where, at
> some places, great numbers of them were taken. Even ants and
> reptiles abandoned, in clear daylight, their subterraneous holes
> in great disorder, many hours before the shocks were felt. Large
> flights of locusts were seen creeping through the streets of Naples
> towards the sea at night before the earthquake. Winged ants took
> refuge during the darkness in the rooms of houses. Some dogs,
> a few minutes before the first shock took place, awoke their
> sleeping masters by barking and pulling them, as if they wished
> to warn them of the impending danger; and several persons
> were thus enabled to save themselves.

Alexander von Humboldt (1822) understood that those people most
fearful of earthquakes "attentively observed the motions of dogs,
goats and swine. The last of these animals, endowed with delicate
olfactory nerves and accustomed to turn up the earth, give warning
of approaching danger by their restlessness and their cries. We shall
not decide whether, placed near the surface of the ground, they are
the first that hear the subterranean noise; or whether their organs
receive the impression of some gaseous emanations which issue from
the earth. We cannot deny the possibility of latter cause."

In China the population is well instructed concerning the observation and reporting of anomalous animal behaviour, which is taken
very seriously as a possible earthquake precursor. The coverage there
of such occurrences is the most complete. The destructive Haicheng
earthquake (magnitude 7.3) occurred on 4 February 1975; reports
of anomalous animal behaviour became frequent as early as
December 1974 and continued to the time of the earthquake itself.

In mid-December, according to one account, "snakes hibernating came out of their burrows and were frozen stiff on the snow. Rats appeared in groups and were so agitated that they did not fear human beings." (It is reported that one person caught 20 rats with his bare hands.) Other anomalous behaviour was observed in domestic birds and animals.

In the beginning of February there was an increase of such reports (Deng *et al.*, 1981), especially among large animals such as horses, pigs, cows, deer and dogs. Rats appeared to behave as though drunk. There were hundreds of such reports. In most cases the kinds of disturbed animal behaviour were regarded as sufficiently unusual to be reported to the seismic prediction network before the earthquake. The locations of the animal behaviour anomalies and of the precursory "earth fog", which hung over the area for a few hours before the quake, both seem to have a close relationship to the line of the fault, which subsequently slipped.

There is little doubt that the animal behaviour stories played a part in the successful prediction of the earthquake and the evacuation of Haicheng, which probably saved tens of thousands of lives. The Chinese seismologists were taking the subject seriously, even if they could not explain it in detail. It can no longer be casually dismissed in the West. The growing body of anecdotal evidence has to be scrutinized seriously.

We have already referred to anomalous behaviour of aquatic animals before earthquakes. A few days before the powerful Kanto earthquake (magnitude 7.9 on the Richter scale) destroyed Tokyo on 1 September 1923, many dead fish were seen in unusually muddy water, large numbers of crabs crawled onto the beach and members of a rare species of deep-sea eel were found floating on the surface. Then, several hours before the quake, some fishermen observed many deep-sea cod floating dead on the surface; fearing a disaster they called off their fishing and returned to land. The Kanto earthquake occurred about noon, but between 7 and 8:00 a.m. that day, according to Rikitaki (1976), many carp came up to the surface of a pond in downtown Tokyo and "it seemed as if these fish suffered from shortage of oxygen. It is said that the fish recovered when they were put in fresh water."

All these animal behaviour effects before an earthquake can be understood in terms of gases issuing from the ground. A large volume of gas coming up from great depth prior to the earthquake will drive before it the normal gases in the porosity of the soil. These are usually rich in carbon dioxide and poor in oxygen. In addition, they contain a great many other molecular components, mostly of biological origin, but some of mineral origin. When this bad air invades the burrows of ground-dwelling animals, they are driven to the surface in order to avoid suffocation. As the gas seeps into the atmosphere it then produces a variety of odours that can be detected by animals with a delicate sense of smell. It presents a kind of chemical "cacophony", having no visible objects of origin, a circumstance that might well be alarming to animals dependent to a large extent on olfactory cues in interpreting their environment and in sensing danger. These two consequences of gas emission – the threat of asphyxiation from below and the presence of strange odours above the ground – were both proposed long ago to explain the precursory animal behaviour.

Other explanations for strange animal behaviour have been advanced (Buskirk *et al.*, 1981), in particular that animals may be sensitive to slight vibrations of the ground, so feeble that people are not aware of them. Many earthquakes are preceded by such weak foreshocks that can be registered only on sensitive seismographs. It would not be easy to understand why animals would respond to such minute vibrations of the ground with the kind of panicky behaviour that is generally described. A truck going along a highway a mile away would create a greater seismic disturbance, and yet not alarm the same animals.

Modern gas-related observations

In addition to all the eye-witness accounts of gas-eruptive phenomena, there are the modern chemical observations which tell essentially the same story. The radioactive gas radon (radon 222) is a product of the radioactive decay of uranium, which is generally present in trace amounts in the ground. After it is produced it only has a very short time before it decays. After 3.8 days, half of it will have decayed. These properties make it an ideal substance for the observation of the movement of gases through the ground.

An individual radon atom, released somewhere in the ground, would only move over distances of centimetres before decaying, if molecular diffusion were the only cause of its movement. If, however, the ground is washed through by another, much more abundant, gas, then of course the radon will be carried with it. For this reason an observation at or near the surface of the radon concentration gives an indication of the movement of ground gases and of their temporal variations. Although from one location to another the amount of uranium in the ground may be substantially different, and therefore variations in radon values from one area to another cannot readily be interpreted, the time variations in the same location tell a meaningful story. If there is an increase in the surface radon, other fluids that are coming up must be responsible.

Radon monitoring has been done in many locations, much of it in the Soviet Union and in China, and a clear relationship to seismic events has been seen. The great earthquake that destroyed much of Tashkent in 1966 was preceded by a gradual rise in radon level over a period of six years, and immediately after the earthquake the local radon level returned back to normal. More commonly, increases in radon level have been seen for much shorter periods, from a month to a few days prior to earthquakes (Hauksson, 1981).

Because of its shortlived and therefore intense radioactivity, very tiny quantities of radon can readily be detected. Gases that can only be detected by chemical means, however, are required in much larger quantity. Helium, hydrogen and methane have all been noted to show similar relationships with earthquakes, in detail different in different parts of the world. Presumably the question of which gases will show themselves prominently, if an outgassing event is in progress, depends on the type of gas content in any one region. Hydrogen and helium have been clearly observed in Japan in this relationship (Sugisaki, 1978; Sugisaki and Sugiura, 1985), and methane in the south central region of the U.S.S.R. (Golubev *et al.*, 1984), and also along the San Andreas Fault (Jones and Drozd, 1983), but in the last case no direct relation with earthquakes has yet been observed.

Another type of evidence which confirms that major gas eruptions accompany earthquakes is to be found in the longlived markings in the ground that they cause. Similar to the formation of mud volcanoes, where there is subterranean mud, massive gas eruptions may expel sand or other subsurface materials that can readily be mobilized. In the great earthquake of New Madrid which we have already mentioned, "sand blows", that is little funnel-shaped sand-filled craters, were created along a stretch of some 300 miles. The eyewitness descriptions of the gas eruption phenomena during the quake are therefore clearly corroborated by these features, which can still be seen today. Moreover it has been possible to discern a set of similar features that are much older and suggest that an earlier earthquake, perhaps some 300 years before, had occurred in the same area.

Other modern observations that seem precursory or related to earthquakes are changes in ground water level, in the electrical conductivity of the subsurface, as well as changes in the velocity of propagation of seismic waves. All these can readily be attributed to movements of fluids through pore spaces and changes of fluid pressures. The water table in the ground may be shifted if gases invade from below, and that, after all, is strongly suggested by the frequent description of water wells changing level, turning muddy or developing a smell. Electrical conductivity is of course very dependent on the disposition of water in the pore spaces and can be expected to be a sensitive indicator of such changes.

"Pockmarks" on the ocean floor

Any markings generated on the sea floor tend to survive a long time in regions of low rate of sedimentation, and of low velocities of water currents. Large areas in which curious circular markings are seen in the ocean mud have recently attracted attention. One prominent area is in the North Sea, where sonar investigations show shallow, circular ridges ranging in diameter, from a few metres to 200 metres over an area of 20,000 square kilometres. The region coincides pretty closely with oil and gas fields of the North sea.

The sonar can see through a certain depth of the ocean bottom mud, and one sees a similar buried region of these pockmarks at a depth of tens of metres. It is estimated that about 10,000 years would have been required to deposit the overburden of sediments. It appears therefore that individual events were responsible for creating a field of pockmarks and that they, like the New Madrid earthquake, were events that released gas at the same time over large areas. One such event must have occurred in the North Sea, perhaps within the last thousand years or so, and another one some 10,000 years earlier.

What we suppose happens underwater in the ocean mud is that individual large bubbles of gas, as they emerge from the mud, pick up some of it and disperse it in the water, leaving roughly circular depressions. This is in contrast to the shapes that are created by a continuous gentle trickle of gas, as is known, for example, in the

Gulf of Mexico, where small, steep-sided cones of mud are generated with the bubbles issuing, as from a miniature volcano, from the top of the cone. Very massive and sudden releases of gas must be responsible for rings that are as large as one or two hundred metres.

Similar shapes have been recognized on the ocean floor in many other parts of the world. They have been reported from the Adriatic, from an area near New Zealand, from the Gulf of Mexico, the Bering Sea, the Great Lakes, the South China Sea, the Baltic, the Aegean, the Gulf of Corinth, the Delta of the Orinoco, and the Scotian Shelf off Nova Scotia (Hovland *et al.*, 1984).

Deep earthquakes: how are they possible?

The depth at which the primary shock of an earthquake occurs can be measured quite accurately by modern seismic methods. Earthquake epicentres occur at all depths from a few kilometres down to as much as 700 kilometres. There is no significant break in that distribution, and we cannot ascribe the deep earthquakes to a totally different phenomenon from the shallow ones. Yet in a theoretical discussion of the fracture and slippage of stressed rock, it would be quite inconceivable that this could take place under the overburden of 700 or even 100 kilometres of rock. The friction between any slipping surfaces under that overburden pressure would be just too great to allow any sudden motion. In fact the friction could resist the shear force just as well as the competent rock before it broke. There just is no critical fracture phenomenon that would suddenly allow the rock to slip and the stored strain energy to be discharged. The elastic rebound theory by itself could only work for shallow earthquakes, and yet the deep earthquakes appear to be no different. Something else has to be at work that allows rocks deep down, and under enormous pressure, to fracture and slip just as they would at a shallow level.

If gases and solids are in a pressure equilibrium, with the gases filling many pore spaces, then this is exactly what would happen. If any developing crack could be immediately invaded by a gas whose pressure were closely matched to that of the rock, then fracture and slipping at great depth would be no different from the process at shallower levels. The gas, by being far more elastic than the rock, would immediately absorb most of the overburden pressure, and the rock would behave more or less as if it had no overburden. It is not quite correct to say that the gas lubricates a fracture and allows it to slip, but it really carries the overburden weight and it eliminates the enormous internal friction in the rock that this would otherwise have created.

Deep earthquakes are confined to particular zones on the Earth, clearly related in definite patterns to the earthquakes at shallower levels. Presumably these are the zones in which gases are ascending and in which the occurrence of earthquakes is therefore facilitated.

Gas emission and elastic rebound

The view that earthquakes are caused simply by stress building up from causes that are unknown, until it exceeds the breaking strength of the rock, seems to dominate most modern research, despite all the evidence we have cited. The fact that the premonitory phenomena that have been identified and that are effective for earthquake prediction do not fit into this simple description is a cause for regret in many publications, but nevertheless there is not much discussion of alternative explanations. We read remarks that "perhaps in China some of these other precursory effects are taken seriously, but we in the U.S. prefer to base ourselves on matters that we understand". It is a strange attitude, since the very effects that are not understood seem to be the ones that work best for prediction.

The main emphasis the U.S. Geologic Survey has placed in its efforts towards earthquake prediction in California is upon the measurement of strain (and tilt, a closely associated phenomenon). A large number of strain and tilt gauges are in place, and in all discussions the strain gauge results are taken as the main indicators of the situation. Yet the results, as far as prediction is concerned, have been most unpromising. The assumption appears to be that all the other phenomena – the radon emission, the ground water level changes, electrical conductivity changes, changes of the seismic velocities, the great range of precursory gas emission phenomena – are only consequences of changes in the strain of the rock. This outlook does not fit the data at all well, and it is not helpful for the development of better earthquake prediction techniques. Once more we see a general reluctance to consider that anything observed in the crust is caused by events below.

The common pattern shown by the precursory effects of major earthquakes is that they are observed a few days or weeks before, over a very wide region that is sometimes a thousand miles or more in dimension. These effects can then be observed for some period, and during that time only quite minor strain changes have been noted; somewhere within the large area that showed the precursory effects the earthquake will then strike, though the precise location cannot usually be predicted.

Could one really believe that a build-up of strain had taken place over the entire large region and had reached some critical level at much the same time, to cause all these other symptoms? Whether those symptoms can be produced by strain is itself very questionable; and how this would be possible without giving a clear reading on the strain gauges is almost inexplicable. But how could one imagine that the build-up of strain occurs in such a way that dimensions of a thousand miles or more reach a similar critical level to give rise to the precursory information within the same week? The time scale of build-up of strain is reckoned in tens or hundreds of years. How could rock masses that are a thousand miles apart, that are subject to different geometrical configurations of stress, that are probably different structurally, chemically, thermally, all reach that critical strain level at the very time when the supposed secondary symptoms appear? We would expect one spot to reach the breaking point first, and then the result of a fracture there would perhaps distribute the stress to other locations. One location would show the critical strain phenomena, break and perhaps cause a larger region around it to break in quick succession, even though no precursors had existed there, but that is not the observed pattern.

I believe that the entire process has to be understood in terms of a combination of strain and gas emission from below. The build-up of strain in different rocks occurs unevenly in different locations and at different depths. In the absence of a pore fluid, the rocks at deeper levels will flow plastically when a critical stress is exceeded, rather than break suddenly. In this case the general tendency will be for rocks to be stressed just up to the level where flow sets in, and higher stress levels would be shed by a virtually imperceptible flow.

If now a gas mass ascends from below, fracturing the rock or opening up old fractures, then the situation will be changed drastically. The generation of many small cracks will weaken the rock; as its ultimate strength decreases, it may now reach the failure point. It is not a sudden increase in the *stress* of the region that brings about the earthquake, but rather a sudden decrease of the *strength* of the rock. Whatever strain energy had accumulated in the rocks can now be discharged.

It is not surprising, then, that little evidence can be found through the measurement of strain preceding a major quake: nothing note-worthy will have happened at the time to the stress field other than the inflation of pore spaces. A rise in the surface level might be measurable, but that is difficult to observe except at the sea shore (and has sometimes been observed as a precursor). Such an inflation may do very little towards introducing horizontal deformations on or near the surface that would be seen on strain gauges, and if the regions that are inflated are large, even tilt observations will not show much. On the other hand, the direct gas-related phenomena may be plainly in evidence over the entire region under which the gas has distributed itself. The leakage of gas to the surface will carry up the radon, it will change ground water levels, it will cause smells and noises, and the change in porosity of the rock will change seismic velocities.

Somewhere in the region affected, the decrease in the ultimate strength of the rock will be the first thing that leads to failure. A fracture will now develop where, without the gas, it would have been merely a flow. Now cracks can open up and propagate, with the gas pressure available to hold the faces apart. What all these precursory phenomena are telling us is: "Beware! The rocks which are normally stressed to near their ultimate strength have now been invaded by a gas. The fracturing that will develop will greatly weaken their ultimate strength and when they give they will do so by way of a sudden violent fracture and not by a continuous flow as was the case before the gas entered".

We can now understand why a large region can give all the precursory information all of a sudden, or at any rate, within the short time of a few days, and why a quake is then likely to occur in some area within that region, without any particular additional premonitory signs there. "We can predict better the time and magni-tude of an earthquake than its location within the large region for which we have a warning": that was the information given out to a U.S. delegation by Chinese seismologists who simply based their predictions on the long-established phenomena, irrespective of any interpretations that have been offered. These were largely the phenomena that can best be understood in terms of the entry of gas into the region as a whole; which point of the weakened region of rock would then reach breaking point first was then a matter of the distribution of stress, for which no detailed measurements existed.

With this general viewpoint we can also understand how large changes of volume of the ground can take place. We have already noted that the large ocean waves, or "tsunamis", require sudden changes in the level of the ocean floor or other changes that displace large volumes of water. On land major earthquakes have sometimes led to the abrupt lowering of some large areas.

An earthquake that occurred on 13 June 1984 near Okinawa seems to have been the first for which a detailed study of a sudden change of volume of the ground below has been carried out. The tsunami it created was not only remarkably large for a modest earthquake (magnitude 5.5 on the Richter scale), but it could be analysed to show that a large extra volume had appeared suddenly, observed by a rise in the ocean level simultaneously over a large area. The authors of this analysis (H. Kanamori *et al.*, 1986) attributed this to the creation of steam, due to a supposed sudden contact that had taken place between seawater and lava. As we shall discuss later, this type of explanation is quite inadequate for many other explosive events for which it has been advanced, and we do not believe it to be applicable in the present case either. Tsunamis of many other earthquakes have also shown such volume changes, or even very much larger ones, only they were not subject to such a detailed study.

The explanation in terms of a local generation of steam seemed the only one available, if one takes the conventional view of denying the possibility of the entry and rapid expansion of a gas from deep levels below.

The old theory of earthquakes as due to the movement and eruption of gases underground and the modern elastic rebound theory thus give together a much better explanation of all the phenomena than either theory did singly, and in particular the numerous precursory effects then have an explanation, and their understanding is the key to the important task of earthquake prediction.

5

OUTGASSING IN SOLID ROCK

Gases can erupt in the most violent explosions from great depth to create the kimberlite pipes, bringing with them carbon in unoxidized form – the diamonds. Volcanoes, which we shall discuss later, bring up large quantities of gases from great depth and the huge volcanic eruptions derive their force from gas pressures at great depth. While the kimberlite pipe eruptions are very rare, volcanoes are active in hundreds of places all over the Earth. It used to be thought that the process of outgassing from deep levels and with that the supply of the volatile substances to the surface of the Earth was essentially carried out by the volcanoes. The solid crust was considered impermeable and firmly shut in most areas most of the time. But is this really so?

In fact there is much evidence that gases do come up in cool regions too. The quantities of helium and other inert gases that appear to be streaming out of the ground greatly exceed what could have evolved from the shallow levels. Major earthquakes, as we have seen, give a strong indication of being associated with the outflow of gas, sometimes over very large regions. The big cracks in the crust of the Earth where earthquakes are most frequent would perhaps make it a little easier for gases to come up and indeed one finds the evidence for gases to be preferentially in such places. Mud volcanoes and other gas eruption phenomena which we shall discuss later are almost invariably placed on the top of major fault lines. Helium and argon are often found in predominantly methane-producing commercial gas wells in such quantity that an origin below the sediments is clearly implied. Outgassing both in the earthquake belts and elsewhere seems to be going on in many places, but where it is nonviolent it is much harder to observe it.

The detailed mechanism of such an outgassing process through solid rock clearly has to be understood. It cannot be simply a process of diffusion of gas molecules through the solid material because such a diffusion would represent a totally insignificant transport even on the long time-scale of geology. Diffusion, we can calculate, may play an important role over dimensions of 100 metres or so, but over larger dimensions it would have to be motion through interconnecting pore spaces by which fluids – liquids or gases – could migrate over substantial distances. Can there be pathways open in the deep rocks for such migration to take place? Let us first consider the overall regime of forces and pressures in the Earth. The deep interior of the Earth, like that of any other large astronomical body, must be a "pressure bath" at each level. Rigidity forces in the solids are very small compared with the weight of the overburden. This, after all, is the reason why the Earth is a sphere. The departures from this rule are what account for the heights of the mountains. The rocks at the base of the mountains have to stand up to shear forces of the order of 3000 bars (1 bar = 1 atmosphere = 1 kilogram weight per square centimetre), equivalent to the pressure forces exerted by their weight. Deeper in the Earth, where the long-term strength of the rock is greatly diminished by heat, the shear strength of the rock, that is the rigidity, would only support a few hundred bars, probably tapering to nothing at a depth of 600 or 700 kilometres.

A "pressure bath" at each level means, more precisely, that the pressures in rock or pore spaces cannot deviate by more than a few hundred to 1000 bars from the pressure given by the weight of the rock overlying a given region. In rock with a density of 3 grams per cubic centimetre (3000 kilograms per cubic metre) the overburden pressure increases by 1000 bars every 3.3 km. At each level, pressures in rock and pores can therefore deviate from one another by the equivalent of only a few kilometres in height. If pore spaces contained a fluid pressure greater than that of the rock, the surrounding rock would be put in tension. Since rock is much weaker in tension than in shear, the pore pressures can in practice be less, but hardly more, than the rock pressures.

It has sometimes been thought that where the rock pressures are very high, all pore spaces would be crushed out, but of course there is no reason for that. Just as a deep-sea fish can exist perfectly happily in the enormous pressures of the deep ocean without having its blood vessels squeezed shut, so veins and fissures at every level in the Earth can co-exist with rock, provided there are no large pressure differentials. Clearly the diamonds have given us evidence of fluid veins at a depth of at least 150 kilometres, but there is no reason why there should not be fluid veins at any level. At each depth the porosity will be simply defined by the proportion of the material that is fluid or solid there. In view of the fact that the mantle is of uneven composition, there may be great differences, both regionally and vertically, in the value of this porosity. Not just molten rock or liquid iron should be considered as filling pore spaces, but also fluids that cannot create a mixture with the rock, because they are chemically immiscible or present locally in a higher abundance than can be mixed into the rock, and such fluids might be nitrogen and helium, methane and carbon dioxide, and perhaps a number of others. Deep down, approximately half-way to the centre of the Earth, the fluid "porosity" jumps up to 100 percent – the liquid core.

Now let us consider the pressure situation on a smaller scale and in detail. If we bore a sufficiently deep hole and don't fill it with any heavy fluid, it will collapse and shut itself. This is just the consequence of the rock not being strong enough, when the large pressure in the rock from the weight of the overburden presses in on the hole with no corresponding outward force available there. In most types of rock, such a collapse would happen somewhere in the depth range between 10,000 feet and 30,000 feet. Most types of rock have a behaviour under stress which is time-dependent in such a way that a sufficiently large force would produce an instant collapse, while a somewhat smaller force would result in a slow creep and eventual collapse. Our hypothetical open hole would collapse instantly at a certain depth, but it would slowly grow shut up to a much shallower level. For geological processes it is the slow collapse that would generally be the important one.

The rocks and sediments on the surface of the Earth are usually somewhat porous, and the pores interconnect, giving the material what is called a "permeability" through which fluids can move. In most areas the fluids – liquids or gases – in those pore spaces are more or less stagnant, or moving only very slowly, and in that case the fluid pressure will increase with depth by an amount simply proportional to the density of the fluid and not dependent on anything else. If the fluid in the pores is water, as is most frequently the case, then the pressure in the pores at any level would just be given by the "head" of water, that is by the vertical height over which the pore space region is filled with water. It does not matter how complex the pathways are; so long as the pathways interconnect, and so long as there are no pressure gradients associated with fluid motion, the pressure is determined by the height and density of the fluid. For pure water the pressure would increase by one atmosphere (one bar) approximately every 10 metres of depth.

What happens then if we have a porous sediment that extends to very great depth and is filled with water? In the pores the pressure increases only by one atmosphere in ten metres, while in the rock it increases typically by about 2½ atmospheres in the same interval, because the rock is approximately 2½ times denser than water. Clearly the pores will collapse at the level at which rock cannot support this pressure difference, just as in the case of the open hole we discussed earlier. If we are dealing with a long-lived geological situation, we must expect that the pores will collapse already at that depth at which the long-term stiffness of rock is insufficient to hold them open. The pore space pressure gradient appropriate to the density of water simply cannot continue down beyond a certain depth, because the rock has a limited strength. At the level at which the rock would yield, albeit at a slow rate only, the pores must close and the continuity of the pore pressure regime must be interrupted.

If fluids had only come from above, as for example the surface water, then one might just think that the porosity domain would end at the critical level at which the rock is crushed shut, and there would simply be no porosity and no fluids in any region below. If, however, there is a supply of fluids from below, the situation is quite different and much more interesting.

Suppose there is a source of gas at some very deep level far below the critical level. If there is enough gas made available, perhaps because the rock heats up and contains volatile components that create a gas pressure, then that rock will be fractured, and a fracture porosity will be generated. (Fluid-pressure fracturing is sometimes done artificially, usually with water as the fluid, and it is then called "hydrofracturing" or "hydrofracking".) The evolved gases at the deep level will thus fracture a certain domain, as soon as the fluid pressure exceeds the pressure in the rock by a small margin. (As we have noted, the material at the tip of a crack is in tension, when the crack contains fluid at a pressure greater than that in the rock. A very small overpressure acting for a long time is then sufficient to create fracture porosity.)

When such a domain of fracture porosity has established a communicating system over some interval of height, then there will again be a difference between the pressure gradient in the fluid and in the rock. The fluid pressure at the top of the domain cannot exceed the rock pressure by more than a small margin, since otherwise more cracks would be created and the porosity volume would be increased until the pressure had dropped to the value which the rock can stably contain. Since the fluid is lighter than the rock, the pressure gradient in it must be less than that in the rock. When the domain spans more than a certain interval of height, the pores at the bottom will then have an insufficient pressure to balance that of the local rock, and they will therefore collapse. In this way a definite limit is established for the height to which a connected pore space domain can grow. If then more gas is supplied at the base of the domain, the system develops an instability; there is no longer a steady solution involving only slow and continuous flow.

The pores at the bottom of the domain having collapsed, the domain is now situated above the source of supply but disconnected from it. The pressure around the supply region will therefore increase again, and fracture porosity will again be created or re-activated. The pressure in this will now be much higher than that in the bottom of the previously created domain, and therefore the moment a connection is established there will be a rapid upward flow of gas. The pressure in the newly created or re-activated fractures will now drop rapidly, and they will therefore collapse. The previously established domain has suffered an addition of fluid and therefore an increase in its pressure. If the top was previously at the limiting pressure, then the addition will now cause fracturing there. New porosity will be created at the top, while the porosity at the bottom collapses. The rule is simply that the moment a domain spans more than the maximum stable height interval, it starts to move upwards. The fluid is what moves; the rock stays in place.

In reality there will of course be many complicating factors to consider. The manner in which the rock breaks would depend on the composition of the rock, its temperature, any external stresses to which it has been subjected, previous patterns of fracturing, and so on. Many quantitative aspects will of course depend on these factors. How much volume will a domain possess before it becomes unstable? Does it spread out sideways, or does it retain a narrow vertical shape?

The maximum vertical height of a domain can be estimated from a knowledge of the density of the rock, the density of the fluid (at the pressure in question) and the strength of the rock under compression. To be exact, one would have to know the time-dependent strength of the rock in detail, and one would have to know the rate at which fluid is supplied, for it is these quantities together that would fix the speed of movement of a domain and its vertical height. For the purpose of a rough estimate, which is all that can be done at this stage, one can take a typical strength of deep crustal rocks for the onset of a slow plastic deformation, and that, together with the densities involved, would define the approximate vertical height

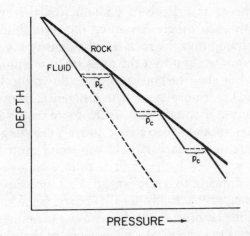

p_c = critical pressure in rock

Fig. 4: A schematic showing pressure regime in rock and fluid-filled pore-spaces. The height of a fluid-filled domain is limited since the rock will support only a limited pressure difference. For fluids less dense than rock, it will fail in compression at the bottom and in tension at the top of a domain that exceeds a certain vertical dimension. (Rock is very weak in tension, hence fluid pressures can hardly exceed rock pressures.) Domains of this critical height will migrate upwards.

of a slowly migrating domain. For hard and cool crystalline rocks, and for the density of methane, one arrives at figures between 4 and 8 kilometres for the vertical height. At deeper levels, where the temperature is higher and the rock therefore weaker, especially for slow creep, the domain height would be smaller (Fig. 4).

If the source of gas we are considering continues to produce, then one domain after another will be launched and will migrate upwards. The separation between domains can be quite different in different circumstances; it will depend on the rate at which gas is liberated, and on the speed of migration of the domains, which in turn depends on the mechanical properties of the rock. One can envision circumstances where successive domains are far apart and separated by a large time interval, but one can also envisage the opposite, where there would only be thin divisions between successive domains. What one cannot have is a continuous open path from a deep source to the surface. If there exists a steady source of gas that has been active for a long time, an almost steady state may be set up. To see what that would look like we have to consider the terminal condition at the surface first.

The pore spaces at the top are open to the atmosphere, and therefore pore fluids and rock meet up there with the same pressure of one atmosphere. At a depth of 5 kilometres, the two will be far out of step with each other. Assuming the pore fluids are mainly water, and assuming that the rock has the density 2.5 times greater than that of water, we find that the pressure in the pore fluid is 500 atmospheres, while the overburden bearing down on the rock is the equivalent of 1250 atmospheres. The difference, 750 atmospheres, or 11,000 pounds per square inch, is the force tending to crush the pores. The softer rocks will give way under a crushing force of this magnitude. Some of the harder rocks can stand up to more, but at a somewhat greater depth they will also have to give way. Experimentally it is difficult to establish at what pressure-difference a given type of rock will actually shut its pores, because it is difficult to reproduce in the laboratory the detailed circumstances of temperature, pressure and the chemical environment of the pore fluids, and it is impossible to observe over time intervals as long as those that will be important for defining the real case.

In practice there will not be an abrupt transition from open pores to completely closed pores. What will happen is that the porosity will decrease with depth, and, depending on the complex statistical properties of the geometry of the pores, there will then be a level at which the communication between pores is effectively interrupted. The long-term strength of the particular type of rock, as well as the geometry of the pores that tend to be created in it, will now determine the depth at which it effectively closes. Immediately underneath that level there can, of course, again be a porosity filled with fluids at a pressure that could be as high as that in the rock, therefore completely stabilizing the rock against collapse. This is exactly what has often been observed by the measurement of pressure and porosity in bore holes, and correlated with that, a decrease in the seismic velocities from above to below this "critical" layer has been seen. Fluid domains that had propagated up from below can then be expected to set up a higher pressure and a higher porosity just underneath the critical layer. Any pore space domain that had travelled upwards and penetrated into the critical layer would have exhausted some of its fluid into the low-pressure zone above. Immediately at such points of penetration the local pore pressure would have decreased to the point where the rock closed again. It is the frictional resistance to the flow that will determine just how quickly and at what level the flow will be pinched off, but it is likely to be such as to leave most of the fluid in the lower domain. That domain will now be effectively arrested, having a little less fluid in it than the amount

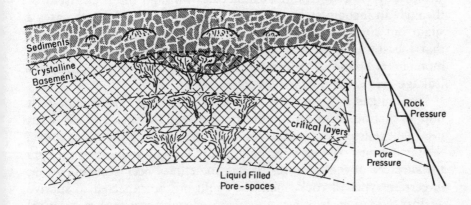

Fig. 5: Fluid-filled domains in a vertical succession of pressure-regimes.
The uppermost critical layer may be in sediments or in basement rock. Low permeability layers in the sediments may provide 'islands' of caprock, retaining a reservoir below, and these were the formations chiefly sought in the search for oil and gas. The critical layers must be continuous surfaces—though possibly often far from flat.

that previously made it marginally unstable to upward migration.

Now if our deep source continues to produce fluids, another domain will have been available to run into the underside of the lower one we have just discussed, and thus a third domain would have become arrested. Eventually domains will stack up vertically, all the way down to the source of the fluids. Once this has happened, any fluid that it added to the lowest domain will cause it to burst through its top for a moment and spill the excess into the next domain above. From here there will in turn be a spillover to the domain above, and so an upward cascade will be set up (Fig. 5).

This upward cascade is the only non-explosive process of outgassing that is possible from deep levels and through solid rock. The pressure gradient in any connected pore spaces being necessarily different from the pressure gradient in the rock, the two can only coexist over a limited interval of height. The fluids must finally set themselves up with a set of pressure gradients that are a stepwise approximation to the pressure gradient in the rock. This can happen only if domains are disconnected from each other. Any time a fluid penetrates from one level to the next it must suffer a large drop in fluid pressure. This can occur either in the case of a momentary rapid flow, through a crack that was torn open, or it can occur steadily by slow diffusion of molecules through the rock itself, or through an incompletely closed-off system of pores. In both cases a slight but continuous evolution of gas at the deep level will assure that the stacked domains will be refilled, either periodically by a momentary spill-over from each to the next, or by a slow, diffusive leakage across the blocked layers.

This outgassing process has several important consequences. It implies that no chemical equilibrium between the rock and the fluids need be expected, because in the domains in which the fluids move, they come into contact with only a very limited amount of rock on the surfaces of the fractures. It has sometimes been concluded that in certain types of rock, methane could not be expected to survive against oxidation. It was assumed in such calculations that chemical equilibrium would be established between a large quantity of rock and a small amount of gas. In the outgassing process discussed here the situation is the reverse – the pathways that can be fractured by the gas will most probably be chemically dominated by it. The amount of oxygen that can be made available by the rock surfaces bordering the gas-filled cracks is limited, and once this is used up no further oxidation will take place along any particular pathway.

Another consequence, and one of great practical importance, is the availability of natural gas underneath the first "critical layer". If in a region there is evidence that methane has been streaming through the surface for long periods of time – and it is possible to have such evidence from various surface chemical indications – then it will be reasonable to expect that methane can be found in a region of expanded porosity and at a pressure near that corresponding to the overburden of rock beneath the first critical layer. Whether the quantities are sufficient to warrant commercial exploitation will depend on the amount and type of porosity existing at that level, and that in turn is a function of the type of rock and the way in which it breaks under gas pressure-induced fracturing. Sediments are generally preferable from this point of view, but metamorphic rocks may be acceptable if they have been subjected to a mechanical disturbance that induced a dense network of fractures in them.

This method of outgassing may provide the explanation for various episodal events to which the crust seems to be subjected. Volcanoes often are dormant for long periods of time and then have a sudden period of violent activity. Could this be due to the penetration of a gas domain into lava channels? Earthquakes show apparently gas-related precursory effects that appear quite suddenly.

Mud volcanoes, like lava volcanoes, suffer sudden phases of activity, sometimes related to earthquakes. The motion of lava or the motion of solid pieces of the crust would always be slow, and it is difficult to see how sudden and sometimes violent phenomena could be produced by either form of motion. Gas motion through solid rock is quite another matter. There very rapid shifts in internal pressure must certainly be expected as fracturing proceeds from one domain to another. Even a perfectly steady and gradual source of gas at a deep level can be expected to result in sudden impulsive pressure variations as the gas ascends. Perhaps all the violent episodal activity of the Earth can be ascribed to the instabilities of such outgassing processes, of which the kimberlite pipes are the most extreme form.

In recent years deeper drilling for oil and gas has resulted quite frequently in the discovery of "overpressured zones". The "normal" pressure in pore spaces in the rock was regarded as that which an overburden of water would have produced. At shallower levels water commonly fills most of the pore spaces to the surface, and therefore this pressure regime is the common one. It is of course out of step with the pressure in the rock, but the rock is strong enough to hold open these pore spaces that are underpressured relative to the rock. In most cases it is a matter of sediments, so that the porosity merely dates back to the original formation process and was never squashed out. It is clear that if we go deep enough we must come either into a region where there are no pore spaces, or into one where there are pore spaces whose pressure is sufficiently high to prevent the rock from collapsing. Any such domain of overpressured fluids must of course be separated from the uppermost domain by an impermeable layer, for otherwise the higher-pressure fluids would merely have escaped upwards until the pressure had dropped to the level at which the pores would close. An impermeable layer would therefore be created in any case. In practice this impermeable layer may coincide with a sedimentary layer that is particularly soft, like clay or salt, or intrinsically dense. But if no such material is present, the critical layer would nevertheless have to establish itself.

The discovery of "overpressured" zones seems to be usually just the penetration of this uppermost critical layer. The novel feature is only that below the critical level there are again pore spaces, and that they are filled with fluids. If one had thought that these fluids had all come from above, or been evolved from the rocks locally, it would require a somewhat unusual set of circumstances to set up this situation. It would have seemed more likely that in the sedimentary sequence of events, each layer would have had its fluids squeezed out as it was sinking to the depth where the rock began to collapse. When it was assumed that there was no supply of fluids from below, special geological circumstances had to be invoked in each area where "overpressure" occurred, and such special circumstances may indeed be present in some cases. But in any region in which fluids have been supplied from below, there must in any case be overpressured regions under the critical level.

In practice, in comparatively soft sediments the critical layer is found to be usually in a depth range between 3 and 6 kilometres. In harder rock it will of course be deeper. Estimates of the long-term strengths of rocks would make these the expected levels.

On the U.S. coast of the Gulf of Mexico, sufficiently many bore holes and seismic investigations have made it possible to map the level of transition from the "normal" to the "overpressured" zone (Jones, 1980). Here one can see that this transition is a surface of many ups and downs, but it is nevertheless a continuous surface over the whole region in which it could be mapped. This is clearly what would be expected if the whole region had fluids supplied from below, but it would be an unlikely circumstance if the source material for these fluids had been above and had been transported down in the sinking rocks.

The uppermost critical layer serves as a caprock which must exist everywhere. If hydrocarbons have come from below, then it will be an effective cap, and generally the last barrier for their ascent to the surface. At shallower levels the fortunate formation of a particularly dense caprock was required to hold down hydrocarbons. At the critical level there is no special requirement. If hydrocarbons are in major part due to outgassing from deep down, then this universal caprock assumes great economic importance. It is underneath this critical layer, the highest one of many that must exist in the crust, that great and accessible reserves of gas can be expected. When new drilling techniques are perfected, it may become possible to reach beneath the next critical layer and into the second "overpressured" zone; but for the present the reserves underneath the first, between 3 and 10 kilometres, will supply the needs.

The drilling engineer who drills a deep borehole into an overpressured zone knows some of these circumstances well. At shallow levels in water-saturated ground and with connecting pore spaces there would be no question that the pressure at any level in his wellbore would be that appropriate to an overburden of water. If he observes any pressure different from this value he must have penetrated through an effectively impervious layer. If there are any such barriers then he may find a pressure higher than that appropriate to a water overburden (the "hydrostatic" value) but at most the value appropriate to the overburden of rock (the "lithostatic" value). This will be the case if fluids from below have penetrated and been arrested by the impervious layer. Alternatively, he may observe an overpressure also if he is drilling through a caprock of limited dimensions, in the shape of an inverted cup in which the pore spaces are filled with gas instead of water. In that case the pressure is defined by the water pressure at the rim of the inverted cup, minus the pressure that the density of the gas would have generated over the height over which it exists. Since the gas is less dense than the water this will result in a value above the hydrostatic one, but usually by a small amount only, since such caps usually do not have a great vertical extent.

The drilling engineer may also encounter an *underpressure* relative to the hydrostatic value, if he has penetrated an effectively impermeable seal underneath which the pore spaces are filled with gas with its correspondingly lower pressure gradient. If this gas has been supplied from below, then the quantities and the long term qualities of the seals will determine whether this body of gas will be overpressured or underpressured (Fig. 6). It is perhaps a likely circumstance that if the seal above is imperfect, the pressure will be balanced within this imperfect sealing layer by the hydrostatic pressure from above. The gas will be pushing against the water in fine pores with a balanced pressure, and although gas is lighter than water the two fluids will not invert their positions, a circumstance which is commonly observed in capillary tubes. The water in the upper domain thus makes an effective contribution to the seal but only so long as the pressure below does not exceed the hydrostatic value at that level. If more gas was supplied from below it would, in that case, not be able to build up to a higher pressure but instead the pressure would stabilize at just that value.

In drilling for gas a departure of the pore pressure from the hydrostatic value in either sense is therefore a positive indication. A gas has to be involved in one way or another to create this situation.

The drilling engineer who drills a deep hole must try to balance the rock pressure sufficiently closely so that the hole does not collapse. He therefore uses a drilling fluid in his hole of an appropriately chosen density. If he is drilling at first through a region where

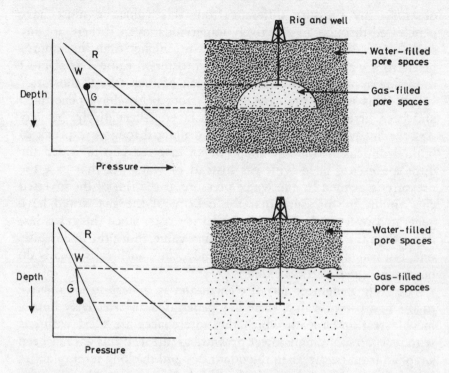

Fig. 6: Schematic showing how 'overpressure' or 'underpressure' can occur where there is a gas reservoir.

'Normal' pressure is taken to be that supplied by water-filled pore spaces open to the surface. R denotes the pressure gradient in the rock, W in water-filled pores and G in gas-filled pores.

the pore pressure has the hydrostatic value, and then through a seal into a region of much higher pressure, the drilling fluid cannot satisfy the requirements in both domains. If he chooses a very heavy drilling fluid so as to match the pressure below the seal, as he might like to do in order to prevent a violent blow-out, then his fluid will exert an excessive pressure on the formation in the upper domain and, if there is any permeability there, it will result in the loss of drilling fluid. He would therefore like to drill first with a fluid of lower density down to the level of the seal, and then case the hole to that level with a pipe, so that he can use a heavier drilling fluid to match an expected higher pressure. Then if he enters a high pressure domain he may have chosen the correct value of the drilling fluid, but only for one particular level within this next region. Since his drilling fluid now has a higher density than any of the fluids that fill those pore spaces – gas, water, or oil – the pressure gradient in his well-

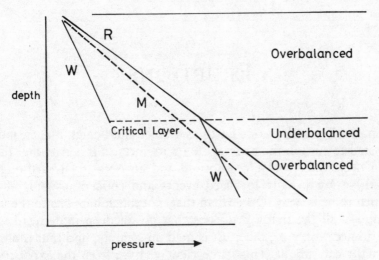

Fig. 7: Drilling with a heavy mud through the critical layer.
R denotes the pressure gradient in the rock, W that in the fluid-filled pores, M that in the well-bore mud, taken to be of a density intermediate between that of rock and water. In this example the well is being drilled 'overbalanced' (relative to the pressure in the pore-spaces) in the first pressure-domain, then 'underbalanced' in the first part of the next domain.

bore cannot stay in step with the surrounding pore pressure. Fig. 7 is a schematic indication of this situation. It is a matter of skill for the drilling engineer to decide on the casing levels and the density of the drilling fluid so as to avoid either a collapse of the hole or an excessive loss of fluid.

In descriptions of deep wells one often finds that a pore pressure gradient appears to have been met which is much higher than the hydrostatic gradient. Since there is no fluid in the pores that is very much heavier than water or water-salt solutions, no such situation can really occur; what happens in such cases is merely the penetration of a layer which is an effective seal, so that there is no continuous pore pressure. Individual pockets of porosity may be encountered in succession and this may give the appearance of a much higher pressure gradient. In any permeable region the gradient cannot fail to be the one given by the density of the fluids filling the pores.*

*Many of the concepts in this chapter are examined in greater detail in a paper by Gold and Soter (1985). Examples and a discussion of pressure profiles in gas-containing basins are given in a paper by T.B. Davis (1984).

6
ERUPTIONS

We normally think of our Earth as a quiet and benign planet, nicely adjusted to sustain the prolific life on its surface. It is true that there is an occasional volcanic eruption, and an occasional earthquake, but these are usually localised events and concern us only if we happen to be nearby. Other than that, not much happens to the firm ground – all the many processes that go on deep underneath our feet do not worry us; the crust is held up steadily, and that is all we normally care about. This is the view we have from the experiences that fall into a human lifetime, or even into the historical record. How does it look when viewed from the perspective of the long time-scale of geology?

On that time-scale the Earth is not nearly so benign. Not only are there internal forces, as yet of unknown origin, that push and squeeze and greatly distort the crust, but more than that, there are occasionally eruptions from unknown causes within, large enough to bring devastation on a continental scale, and quite possibly sometimes large enough to bring global devastation to the fragile biological systems, through changes in climate and atmospheric composition.

There has been a great deal of discussion in recent times about the disasters that may have occurred occasionally and caused the mass extinctions of many forms of life with which the geological record seems to be punctuated. Some think that these extinctions may have been the result of impacts of giant meteorites – asteroids which were large enough to make longlived or permanent changes in the atmosphere. No doubt such impacts have happened and are responsible for some global disasters, but there is every indication that eruptions of internal origin have been devastating on a global scale also.

Most people know of the great volcanic eruption of Krakatau, a small island between Java and Sumatra. It was the biggest eruption in historical times of which we have detailed reports. On 27 August 1883, there began a set of explosions which ended in a violent blast that could be registered in the atmosphere the world over. Seawaves killed some 30,000 people on the neighbouring islands. Volcanic dust and sulphuric acid droplets floated in the upper atmosphere and depressed the mean annual temperature. It was a most impressive event, but even so it was only minor when compared with the really big ones that have left their record. The estimates for Krakatau are that it ejected 20 cubic kilometres of ash, or something of the order of 20 or 30 billion tons. It is an impressive quantity, but small even when compared with another Indonesian eruption in the same century, namely that of Tambora, in 1815. For that the estimate is of 150 to 180 cubic kilometres of ejected ash, or about six times as much. In one brief explosion more material was moved than was ever moved in all the mining operations the world over. Global temperatures were depressed and 1816 was called "the year without a summer". It was in that summer that Byron wrote the poem "Darkness": "I had a dream, which was not all a dream, the bright sun was extinguished. . . ." Quite possibly he was moved to this by the sinister climate of that year, caused by events about which he could not have known anything.

But even Tambora does not measure up to the biggest events that occur on a hundred thousand year time-scale. The Toba eruption in Sumatra, which occurred about seventy-five thousand years ago, produced something of the order of two thousand cubic kilometres of ash (one hundred times more than Krakatau) distributed over the Indian Ocean and most of the Indian subcontinent (Ninkovich *et al.*, 1978).

There were many others that have left the record of deposited ash. The eruption of Santorini, in the Aegean Sea, is the best known because it is thought to have wiped out the civilization in Crete three thousand five hundred years ago. Other records of eruptions in the last 100,000 years that have deposited some tens of cubic kilometres of ash exist from Alaska, Central America, the Caribbean, Japan, Italy, New Zealand. It clearly is a widespread phenomenon. Events in earlier times are more difficult to recognize, but there is no reason to doubt that equally large or larger eruptions have been the rule in most of geological time. Quite possibly some of those were responsible for the mass extinctions of plants and animals, and through long-term effects on the composition of the atmosphere and the climate, they may also have brought about the periods of rapid transition of many forms of life that have been recognized in the geological record. What goes on deep in the Earth to cause such violence?

When the eruption of Krakatau was first discussed, many people thought that it was due to seawater having broken into a lava chamber, where it was turned rapidly into steam, causing the explosions. It was a theory that could be faulted on several grounds, the most important one being the slow mixing and heat transfer between lava and water. Contact between the two is always separated by a layer of steam, and an intimate mixing on a time-scale of less than a minute between huge quantities of water and lava cannot really be envisaged. But, in any case, now that we know that Krakatau was only a small event among a set of much larger ones, we clearly need an explanation that could cope with them all. Water running into a lava chamber certainly could not.

Other explanations have been put forward that depend on dissolved gases in the magma coming out of solution as the pressure is lowered. Convection in the magma chamber has been discussed as a way of lowering the pressure in some part of the magma and this was then thought to lead to a progressive depressurization and explosion of gas. It is difficult to see that any such process would really make violent, sudden explosions, with periods of calm in between.

Another view that has been mentioned is that "enormous pressures" of gas build up underneath the volcano, and finally rupture the rock to burst out, causing the sudden explosive event. When Mount St Helens grew a big bulge prior to the explosion that removed it and a major part of the mountain, this seemed to confirm such a theory.

But this viewpoint is also quite unrealistic. Rock on a large scale has no tensile strength. There is no way in which a volcano can act like a steam-boiler, which explodes when the pressure inside builds up to an excessive value. Rock on a large scale is always full of cracks, and a slight overpressure would immediately open them up. In a static condition the pressure in a lava chamber must always be very close to that given by the weight of the overburden. When Mount St Helens grew a bulge, it must have been due to a rise in the lava level, with the rock blanket that made up the bulge just being borne up by this lava. The violent explosion that followed has to be explained by a very sudden rise in gas pressure in or under the mountain.

With so much evidence of massive gas eruptions having taken place, from the kimberlite pipes to the giant volcanic eruptions, it is strange that the simple picture of the ascent of gases from great depth is not considered in any of the explanations. Why should one consider gases dissolved in the lava just being brought up with the lava and exsolving near the top, but not consider free gas coming up from below? Why should one think that at deep levels the amounts of gas that were available were always only at a concentration below that which could be dissolved in the lava? It is not stated in these discussions, but it seems to be implied, that the authors think it impossible for free gas to exist at depth in the Earth.

Presumably this restriction of acceptable assumptions still stems from the time when it was thought that the entire Earth had once been a ball of liquid rock. In that case no free gas could have been retained in the interior, and there would have been gas retention only up to the solubility level in the liquid rock. In the light of the modern view, that the Earth accumulated from solids, there is every expectation that free gas would be generated when these diverse solids were heated. There is no reason therefore to restrict the discussion to gases coming out of solution in the lavas.

If free gases are indeed generated, and are available in the mantle, then there is little difficulty in accounting for violent, explosive phenomena. We have discussed how gases can ascend through solid rock, either explosively, as in the kimberlite pipes, or gradually, in the structure of migrating domains. In volcanic regions, where liquid rock is present in large quantities, the ascent of gases can take a different form.

The density of lava generally differs only by a few percent from the density of the solid rock. For this reason, lava channels can span a very large interval of height, much larger than the height that can be spanned by a gas domain. While the strength of rock limits a gas domain to a height interval of between three and ten kilometres, a lava-filled region can span the entire thickness of the crust, of between twenty and sixty kilometres, and it can penetrate even to some depth into the mantle. Lava channels therefore hold open pathways where a mass of gas can make a very rapid ascent from great depth.

In the discussion of the ascent of gas in solid rock, we concluded that this has to occur in the form of domains of porosity, migrating upwards by the process of hydraulic fracturing or reactivation of faults in the rock. If this takes place in the vicinity of the magma channels of a volcanic region, there is a chance that a domain will suddenly break into a magma channel and deliver a mass of gas at that point. The immediate effect must be to displace magma, and

therefore to raise the surface of the lava in the volcano. A sudden
lava eruption may be set in motion.

So long as we consider merely the processes involving magma and
solid rock, it would be reasonable to predict only slow changes of
the lava level, dictated by the speeds at which strains in the crust
change. Such variations of strain would alter the volumes of the
magma chambers, resulting in a gradual rise or fall of the lava
surface. This behaviour is certainly common, but in this picture we
see nothing that would make the characteristic sudden onset of
violent volcanic activity, usually beginning with a rapid rise of the
lava level and then sometimes followed by the explosive events.

Gas fracturing its way into a magma chamber provides the reason
for a sudden onset, but is also a good explanation of what is to
follow. A gas bubble – or probably more accurately a volume of
magma foam – may reach the ducts that go up more or less vertically
towards the surface. The gas bubble has a large buoyancy force, and
it will therefore begin to race upwards. It will expand rapidly, since
the pressure of the surrounding magma decreases by one atmosphere
every four metres. Magma will therefore continue to be displaced,
while the gas bubble races upwards, and lava will presumably pour
out of the volcano during this brief time. The increase of volume of
the gas, of course, accelerates greatly (ideally it will double its volume
every time it halves the depth), and so as it gets near the top the
expansion becomes very violent and the eruption of overlying lava
extremely fast. At some level fairly near the top, the expansion of
the bubble will be so rapid that the overlying lava can no longer
form a continuous cap. Instead, an unstable flow will set in: fingers
of the gas will tear through the liquid, and violent turbulent mixing
will take place, and this mix will exit explosively from the volcanic
vent. The enormous quantities of ash that are shot out from the big
eruptions could only have been produced in such a phase of turbulent
mixing of gas and lava.

If we want to think about the gas dynamics and the energy of the
explosion in some detail, we have to recognize that the gas will not
cool much, although it will have expanded in volume by a very large
factor, along with the decrease in pressure (from a pressure in the
mantle of perhaps more than 10,000 atmospheres to one atmosphere
at the surface). It is the close contact with a large mass of hot magma
at each level that will keep the temperature nearly constant despite
the expansion. The bulk of the energy for the explosion is then really
derived from the heat content of a large mass of magma.

Modern drilling engineers, who drill deep wells to produce gas from highly pressured zones, know this type of phenomenon well. They have learned to recognize that a sudden, small rise of the drilling fluid (frequently a heavy "mud") can be a danger signal. If they do not shut off the well-bore firmly at the top before too much mud has been expelled, a violent explosion may follow. It is of course just the same story as the one we discussed for the volcanoes. A gas bubble from below races upwards, expanding enormously and therefore expelling more and more mud, but the speed of its expansion becomes explosive when it expels the last few metres of drilling mud and expands freely into the atmosphere.

If the drilling engineer takes note of the initial danger signal and closes off the well near the top, a gas bubble will of course still rise, but if the bulk of the drilling mud is maintained, its weight will continue to counteract a rapid inflow of gas at the bottom and the gas bubble that does arrive near the top will not be free to make the last violent expansion down to atmospheric pressure. Modern drilling engineers know what they have to do, but unfortunately volcanoes are not equipped with blowout preventers, and the best one can do when one sees a sudden outpouring of lava is to run.

When a violent eruption is over, very often a huge collapse of the surrounding rock takes place, leaving a big volcanic caldera. This is a crater, vastly larger than the hole that forms the vent on the top of a volcanic mountain. Calderas are large, more or less circular, depressions, and they represent the collapse of the overlying and surrounding rock, which suddenly finds itself unsupported when a huge volume of lava has suddenly been excavated. The volume represented by this collapse probably is approximately equal to the volume of lava expelled in the eruption.

What is the composition of the gases that come up in the eruptions? There are very many measurements of gas composition in volcanic regions in fairly quiet times. Water vapour and carbon dioxide are generally the dominant gases but a few percent methane and hydrogen are often recorded also. But is this representative of the large amounts of gas that come out in violent eruptions? In the slow transport of gases through magma, when the gases are in solution or in the form of small bubbles, they will form a chemical equilibrium with the magma. If the magma provides oxygen for the oxidation of methane or hydrogen, then this will be converted largely to carbon dioxide and water. The chemical composition of the magma will determine the oxygen availability and the amount of the unoxidized gases that might remain.

In a violent eruption the situation may be quite different. The quantity of gas racing through the magma may be so large that it dominates the chemistry of the magma and gas mix that is produced. In this case the amount of oxygen that the magma can provide may be quite insufficient to oxidize all the gases, and methane and hydrogen may survive in much larger proportion. Of course it is difficult to sample the gases at the volcano at the time of a major eruption, but if very large quantities are involved it may be possible to recognize the gases as an addition in the mixed stratosphere some time later. (Unexplained increases in the atmospheric concentration of methane have been seen, but so far it has not been possible to relate this phenomenon in detail to major volcanic eruptions.)

It is of interest, however, that there are several reports of flames seen at or around the times of volcanic outbursts. Small blue flames coming out of a frothy lava as it rolls down a mountainside seem to be a common occurrence in Iceland and Hawaii. Flaming lava has also been observed in Africa. There are several reports of large flames shooting out of a volcanic vent during an eruption, but perhaps one cannot be quite sure of that identification; hot luminous sprays of ashes may give a similar appearance. The Tambora eruption in 1815 had such descriptions of flames (Raffles, 1817), as did an eruption of Volcan Fuego in Guatemala in 1974 (Dawson, 1981).

A much more definitive description of flames is available from an eruption of Krakatau in 1928. There, when for some days there were volcanic discharges underwater, orange-yellow flames were reported as dancing on the surface of the water: "The entire surface of the water above the crater was like a sea of flames. Seen from a distance of about 200 metres the flames were about 10 metres high" (Stehn, 1929). (The colour of the flames in this case is probably dominated by sodium picked up by the gas in the seawater.) Here there can be no confusion with incandescent cinders. Flames were also observed during the eruptions of Santorini in 1866 and of Pelée in 1902 (Lacroix, 1904).

There is much evidence that gases of quite different composition come up from the mantle in different areas. There probably is no such thing as a particular mix characteristic of volcanic eruptions. Different volcanoes, and possibly the same volcanoes at different times, may produce different mixes, and then in turn these gases may suffer different degrees of chemical change, such as oxidation or dissociation, depending on the speed of their flow through the magma.

The chemical make-up of the gases at volcanic eruptions makes a big difference to the amount of devastation that is caused. If carbon dioxide is a major component, then the emerging gas will tend to be heavy and, if it is not very hot, it may be heavier than air. In a hot volcanic eruption some gas masses may also be heavier than air just because fine particles of volcanic ash may be mixed in with the gas. In all those cases much devastation is caused by the heavy cloud flowing down the mountain sides and collecting in the low spots. The ash-loaded gas flows bear the name of "nuée ardente", which has been the most terrible of volcanic phenomena in historical times. Such an eruption of Mont Pelée in 1902 on the island of Martinique totally destroyed the city of St Pierre, killing all its 28,000 inhabitants.

An emission of carbon dioxide at a low temperature can also be a devastating event. Where the gas comes up in a volcanic region, but through a vent which is not hot at the top, it may have been largely oxidized in the lava below and then cooled in the ascent. A huge mass of carbon dioxide may then roll out over the countryside and asphyxiate people and animals. To add to the problem, partly oxidized volcanic gases often contain hydrogen-sulphide, which is extremely poisonous, and the mixture of carbon dioxide and hydrogen-sulphide spreading out over the ground is the worst of all. It was an eruption of this kind, with gases bubbling out of a lake that had formed in a volcanic crater (Lake Nyos), which killed more than 1700 persons in a remote volcanic area of the Cameroons in August 1986. In Cameroon another similar event took place two years earlier at Lake Monoun, killing 34 persons. This type of phenomenon is not restricted to volcanic lakes in Cameroon. An eruption of heavy and poisonous gases took place in Java in 1979, killing 142 persons (Le Guern, 1982). In this case the gases emerged apparently from a fissure on high ground and flowed down across a lower-lying village. A later measurement in the fissure showed that carbon dioxide and hydrogen sulphide were present.

Mud volcanoes

There is another eruptive phenomenon, much less well-known but very interesting from our point of view, and that is the phenomenon

of mud volcanoes, mentioned earlier. When they occur it is not lava
and gas, but mud and gas that are brought to the surface. (Higgins
and Saunders, 1974; Ali-Zade *et al.*, 1984). The mud is simply a
sediment, well mixed with water so as to be a thick, viscous and
dense fluid. Where such mud has poured out of a vent for a long
time it has generally built up a volcanic cone as it dried on the
surface and became rigid. Such volcanic cones look remarkably
similar to the cones of lava volcanoes, but are not quite as large.
Still, they are sometimes of an impressive size, several kilometres
wide at the base and several hundred metres high.

Mud volcanoes are similar to lava volcanoes also in the way in
which eruptions occur. Often the mud level rises and an outpouring
begins, sometimes followed by a gas eruption. The gas eruptions can
be very violent, though not quite matching the violence of lava
volcanoes. In the great majority of mud volcanoes the emerging
gases are combustible and consist mostly of methane, though other
components of hydrocarbon gases are also present. When the gases
emerge at a high speed, they generally ignite, presumably from
electric sparks which occur there (just as in lava volcanoes, where
lightning is produced when particles carried in the stream transport
frictionally created electric charge). Some flames of mud volcanoes
have been quite impressive. A mud volcano erupting near Baku, on
the shores of the Caspian Sea, had a flame that shot up to a height
of 2 kilometres (verified by a photograph taken from Baku) and
then burnt to a lesser height for 8 hours. The orifice from which the
gases came had a diameter of 120 metres (Sokolov *et al.*, 1968).
Clearly a very large quantity of gas was involved in this eruption,
as in others that occur in the same area on a time-scale of 10 years
or so. One can estimate that an eruption of this magnitude required
something of the order of a million tons of gas.

In several other respects also the mud volcanoes parallel the behaviour of lava volcanoes. Eruptions are often, but not always associated
with earthquakes. Mud volcanoes generally lie over faultlines, and
in fact all the major mud volcano regions of the world fit neatly on
the global map of major crustal faults (Fig. 8). The largest mud
volcano region, north of the Caucasus mountains and on the shores
of the Caspian, shows the lines on which mud volcanoes occur
closely following the underlying faultlines in the basement rock.

What then can be the detailed explanation of the mud volcano
phenomenon? It is rare for large quantities of fluid mud to exist
underground, and yet the quantities of dried up mud in some mud

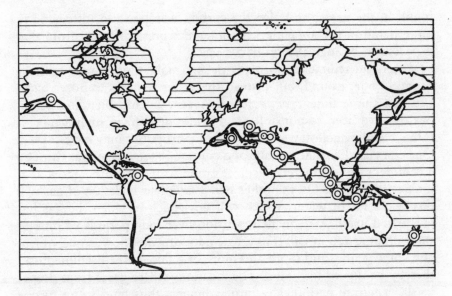

Fig. 8: Black lines show geologically young (mesozoic and tertiary) faults and fold belts. The 13 most prominent mud volcano areas are marked, and they all fall on these belts.

volcano regions point to huge amounts that have been excavated, in many cases over periods of several million years. The quantities of gas associated with eruptions in the Soviet mud volcano fields have been estimated to be so large that any known gas field would have been depleted in a small fraction of the age of the mud volcanoes. Probably the same is true in many other mud volcano areas. Why are there unusually large quantities of mud underground in just the places where also large amounts of gas make their exit?

One theory has been that the mud generates the gas. This view-point is hardly tenable on quantitative grounds, and it also could not account for the great violence of some of the eruptions. Gas being generated within the organic content of the mud by bacteria would be bubbling up through the mud on a continuous basis, and it would be hard to see why large quantities could suddenly emerge with violent explosive force.

Chemical investigations also seem to show a different origin. The methane that emerges often has significant quantities of ethane and propane and other hydrocarbons with it, while bacterially-produced methane in the mud would be expected to be free from such components. Mercury, helium and other trace constituents have been found in the gases, and also the carbon isotope ratio (discussed in more detail in the appendix) is sometimes quite different from that expected in biologically produced gas.

Water and sediments normally do not make the intimate mixture of mud underground, but instead the water remains in pore spaces between the sedimentary material. A very intense blending action has to have been responsible to generate mud from such a mix. It is this blending action which periodic gas eruptions may have caused. One can visualize that a gentle and continuous stream of gas through a sediment would merely displace some water in pore spaces as it makes its way up. On the other hand, a violent injection of gas will displace a lot of water, which will flow back in when the gas has exhausted itself. A frequent repetition over long periods of time of this pumping action will thus create a lot of water flow, some violent and some gentle, and more and more of a fine-grained sediment can thereby be turned into mud. The locations where gases from deeper levels penetrate periodically into sediments thus tend to be turned into underground mud reservoirs. The same action of the gas will then cause the mud eruptions and build the structures of mud volcanoes.

In lava volcanoes, magma channels can span great intervals of height because the magma is almost as dense as the rock. Mud is of course less dense than rock, but on the other hand much denser than water. For this reason mud reservoirs and deep mud channels can be set up that span larger vertical intervals in continuous channels than would be the case for water, gas or oil-filled pore spaces. When an injection of gas occurs into a mud channel it may therefore allow a gas bubble to ascend a long way, and to expand greatly, producing the explosive eruption. Since the channels cannot be quite as deep as magma channels, and since the pressure does not change as rapidly with depth in the mud as in the magma, the resulting explosions cannot be quite as violent. Also the gases are cool, and for that reason the energy content is much lower. Nevertheless, the two types of eruption show many similarities because they are probably both caused by the same type of process, namely the sudden ascent of gas injected at depth and coming up through a long, fluid-filled channel.

We have noted that most mud volcanoes emit mostly methane, while lava volcanoes emit mainly water vapour and carbon dioxide. Can one suspect that similar gases enter from below and that the difference arises only because of subsequent chemical action?

Of course the steam would in any case be absent from mud volcanoes, since the mud is cool, and even if steam were available at depth it would be condensed on the way up. If methane were the chief carbon-bearing gas injected into a lava volcano, it would be turned largely into carbon dioxide, with oxygen available in the magma, except perhaps at times of very violent eruptions. It is interesting to note that there are a few mud volcano regions, the most notable one in Alaska, where the mud volcanoes are close to lava volcanoes, and where the emerging gases are largely carbon dioxide. These observations tend to confirm the viewpoint that the deep carbon-bearing gases in strongly outgassing regions are composed largely of methane and other hydrocarbons at depth, are oxidized if they are in contact with magmas at shallower levels, but emerge essentially unaltered in cooler exit channels.

All the evidence of eruptions therefore shows that large quantities of some gases accumulate at depth and can blow out with violence. This can occur either through vents that are created by the high pressure gas itself, or through vents that are held open by another fluid. These gases are frequently flammable.

The sudden onset of volcanic eruptions can be understood best as due to a very rapid flow of gas through a freshly opened crack in solid rock, leading into a magma channel.

7

THE EVIDENCE FROM HELIUM

The noble gases, of which helium is the lightest, can give a great deal of evidence about the outgassing processes that are taking place, and they can also give some indications about the cumulative effect of outgassing in the past. Helium is particularly useful for the study of the ongoing processes for one important reason: it does not accumulate indefinitely in the atmosphere, but is lost from the Earth into space. Outgassing of helium can therefore be identified rather readily, since many sources of gas in the ground are greatly enriched in helium compared with the atmosphere. The quantities coming out of the ground are orders of magnitude more than those of the other noble gases. The particular importance for a discussion of the outgassing of methane stems from the fact that there clearly is a close relationship between methane and helium in many parts of the world.

As a noble gas, helium has no part in any biological processes, and a biological concentration of helium can be ruled out. It would be difficult to see why helium and biologically-produced gases should frequently share the same reservoir. The association of methane with helium is one very simple and clear fact that raises doubts about the biological origin of methane. But let us look in more detail into the helium story.

Most helium on Earth has resulted from the radioactive decay of uranium and thorium. The particles emitted in this decay are alpha-particles, the nuclei of the helium-4 atom. Once brought to rest they pick up the two electrons needed to form a helium atom, an atom of a chemically quite inert gas. In the atmosphere helium is present at a concentration of 5.24 parts per million by volume. But helium has another stable isotope, helium-3, which is much less abundant. In the atmosphere there is less than $1\frac{1}{2}$ parts helium-3 to a million parts helium-4. (The precise ratio is 1.4×10^{-6}.) In the surface of the Sun and in the solar wind that blows out from it, one observes a ratio of helium-3 to helium-4 of 3×10^{-4}, or about 200 times higher than the ratio in our atmosphere.

The primordial helium that was incorporated in the Earth, as a small impurity in the rocky grains, must have been of solar composition, and thus much richer in helium-3 than any present-day terrestrial helium. But as the total quantity that was brought in by the grains was so very small, and as the Earth in its formation process seems not to have acquired any free gases, the continuous production of helium from the radioactive decay of uranium and thorium, during the whole age of the Earth, became a major factor. This decay produces essentially only helium-4, and as this contribution grew steadily, the average ratio of terrestrial helium tipped heavily in favour of helium-4 in the course of time.

While a small proportion of helium-3 can also be produced by radioactivity in an interaction involving uranium and lithium, this is not thought to be a major contribution to the helium-3 that the Earth now has. The main source appears to be the primordial helium-3 that was imbedded in the solid particles that formed the Earth.

There is no way to distinguish the helium-3 generated by the lithium reaction from the primordial helium-3. However, since the reaction is a small by-product of the radioactive decay process that necessarily produces helium-4 in much larger quantity, it is clear that the process cannot produce a ratio of helium-3 to helium-4 greater than a certain value. In the most ideal circumstances of production, where lithium and uranium are in close association, the production ratio of helium-3 to helium-4 can be as high as 1.2×10^{-5}, or 10 times the ratio in the atmosphere (Aldrich and Nier, 1948). But this only occurs in a rare mineral, and the average production in the crust of the Earth of helium-3 by such processes has been estimated as at least a hundred times lower. This means that in any location where we find a helium-3 to helium-4 ratio of more than one-tenth the atmospheric value, there is the suspicion that we are looking at primordial helium outgassing, and not at local production. Certainly when the ratio is equal to the atmospheric one, it is exceedingly unlikely that local production would be involved. (One must of course be sure that no atmospheric contamination is involved, and one can do this either on the grounds that the total helium concentration is too high, or by observing other noble gases and their isotopes, which have different abundances in the atmosphere, as in the interior of the Earth.)

The atmospheric ratio of helium-3 to helium-4 may be to some degree unrepresentative of the mean ratio that is now coming out of the Earth, since helium-3 may be expected to evaporate more readily into space and thus be depleted a little faster. How large this effect is in practice is not completely clear; it depends on the patchiness of the temperature distribution of the outer atmosphere. A very patchy or fluctuating temperature distribution will tend to minimize the differences in the evaporation rates of helium-3 and helium-4.

What then is the evidence that these two gases can give us? Where helium-3 is above the concentration that can be ascribed to radioactive production, we know that we are dealing with a gas that could not have become chemically concentrated on the Earth in any way. The only concentration it could have suffered is due to a flow that has taken the atoms from their sparse distribution in the original rocks, through a system of pores in which the gas could flow, or through liquid rock in which it could bubble, to the locations where we now find a high concentration. Only mechanical funnelling from a large internal volume can concentrate this gas, and it is therefore an excellent indicator for finding major outgassing locations on the surface.

Uranium and thorium are in a much higher concentration in the crust of the Earth than in the mantle rocks, and therefore much of the helium-4 is generally considered to be produced in the crust. One has thus come to regard a low ratio of helium-3 to helium-4 as an indication of a crustal gas and a high ratio as an indication of a mantle gas (Lupton, 1983).

The need for carrier fluids

But it may not be as simple as that. Even in rock of high uranium and thorium content, the helium produced will never reach a sufficient concentration for the gas pressure to cause fracturing of the rock and to set up interconnecting pathways. Without that the transport would only be by diffusion in solids, and even though the helium is more mobile than any other gas, a general outgassing of crustal rocks by diffusion is a negligible process. In the entire age of the Earth diffusion would transport the gas only over distances of a few hundred metres.

Not only would diffusion of individual helium atoms in the rocks be so extremely slow as to be unimportant for long-distance transport, but it could never account for any significant concentration of helium in a volume of rock or pore spaces. Diffusion processes of individual atoms are hardly affected by gravity; therefore no significant tendency for an upward motion would exist. Diffusion will merely distribute the helium gradually further and further away from its source region, and the effective driving force would always be in the sense of decreasing the spatial concentration. It can therefore not be argued that helium would become concentrated under a caprock if it were transported by diffusion. The concentrations of helium found in practice, in helium-rich gas reservoirs, are many orders of magnitude greater than the concentrations that would result from any uranium and thorium deposits in the sediments, or anywhere else.

The comparison has often been made of the quantity of helium mixed with the gas of a gas field, and the total quantity of helium that could have been produced by the radioactivity of the entire sedimentary basin. In many cases (as the one we shall discuss here) there is far too much helium in the gas field, even if every atom produced in those sediments had been transported under the caprock of the gas field. But in fact no such concentration of helium would occur. Diffusion would take the helium to a lower and lower concentration, whether this diffusion was in the rock itself or in the pore spaces, or even in the layer regarded as a caprock of the gas field. If diffusion alone were at work to rearrange the disposition of helium, then the concentrations that we could expect in gas fields would be thousands of times lower than those that we find, and no enhancements of the helium concentration could be expected anywhere except near source materials from which it had been produced.

How then can the large concentrations of helium that we find in the pore spaces of rocks have been produced? Helium concentrations of about 1 to 3 percent occur in many gas fields of the world, and in a few cases concentrations as high as 9 percent have been recorded. This is in connected pore spaces that make up perhaps 15 percent of the volume of rock, implying that of the order of one atom in a thousand in such a locality is a helium atom. All these high concentrations occur in locations where methane, as well as other components of petroleum, and frequently nitrogen, are caught under a layer dense enough to prevent their upward motion and escape into the atmosphere. Unlike the case of diffusion, we are dealing here with the bulk motion of gases, driven by the buoyant forces of gravity in a major way. In any one locality the gases tend to move upwards, as heavier fluids like water and oil compete with them for the available pore spaces. On a larger scale the fact that the gases are lighter than rock will by itself cause upward mobility, as we have discussed. In each case the bulk gas can be driven against a barrier and concentrated there.

This explains how the force of gravity can concentrate a gas-mix under a high spot of a caprock in much the same way as gravity will concentrate water in a lake at a low spot of the surface. Only bulk flow, subject to the forces of gravity, can do this, while diffusion could not. But this does not yet explain where and how the proportion of helium to the other gases can reach the observed very high values.

As we have noted, bulk motion of gases through pore spaces can only occur where there are sources of gases or liquids that are sufficiently concentrated to force open fractures in the rock and to hold them open against the rock pressures. When such pathways of pore spaces have been generated, then the sparsely and diffusely distributed helium that is liberated in the rocks, within distances that can be covered by diffusion, can join the stream of the other fluids. If these pathways are long, much helium may be added to the stream, and it would be perfectly possible in the end for the quantity of helium gathered up to be quite comparable with that of the other gases or liquids; nevertheless the other fluids were necessary to liberate the helium. (A more complete discussion of helium concentration will involve the partial pressure of helium within the rocks at the ambient pressure, which diffusion will tend to equalize with the helium partial pressure in gas-filled pore spaces. A high ambient rock pressure, and hence a great depth, will favour setting up high helium concentrations.)

Without these conditions, a diffusely distributed gas is permanently imprisoned in the rock. It is only when a fluid is present which, at its source, can be produced in so high a concentration that it can create the fractures, that the possibility of large-scale motion arises. This means that regions with a high content of radioactive minerals are not necessarily regions of a high helium flux to the surface. The existence of a more abundant carrier gas, and the pathways it has created, will be a far more important factor in determining the outflow of helium in any region, than the variations in the concentration of the radioactive elements. Many attempts at prospecting for uranium ores by searching for helium have failed for this reason, while such a search often gives a good indication of the presence of methane or oil, apparently because these substances are frequently associated with an upward flow of gases that become carriers for the diffusely generated helium.

With this picture in mind, the ratio of helium-3 to helium-4 in a reservoir or a vent may have quite a different meaning from the usual assumption that it allows one to distinguish crustal helium from mantle helium. If a given pathway, made and held open by an abundant gas, has existed for a long time, the neighbourhood of all its crack-surfaces will have become depleted of its original helium-3; the carrier gas will no longer bring up much of it. Helium-4, on the other hand, is still currently produced along those same pathways, and it may therefore be brought up even in pathways that have been active for a long time. In that case the helium-3 to helium-4 ratio would be more indicative of the age of a pathway than of any other feature. A young pathway in the mantle rocks will bring up a helium-3 rich gas, while an old one, which has been swept by the carrier gas for a long time, will produce a low helium-3 value.

While one cannot envision any rich helium-3 source in the crust, one certainly could have helium-4, with a low proportion of helium-3, coming from old pathways from the mantle. A high helium-3 ratio (greater than atmospheric) will therefore certify a gas as derived from the mantle, but a low value will not certify the gas as deriving from the crust; it may be crustal, but it may also be from an old, deep source. High concentrations of total helium, of any isotope ratio, can probably arise only from deep sources of the carrier gas, which would sweep it up on the long pathways towards the surface.

With this explanation we can understand a feature in the occurrence of helium which it is difficult to account for in any other way. It is the fact that there are regions on the surface of the Earth with dimensions of some hundreds of miles, where helium has a higher-than-usual average abundance. This may manifest itself in the helium proportion in wells, in ground water, and even in the surface soil. It seems unconvincing to argue that in such regions two independent sources of helium have been responsible for making it helium-rich. Yet it often occurs that the helium-3 to helium-4 ratio is widely variable within the zone. Surely we would not think that such a region was outstanding for its supply of mantle gases, and independently also outstanding for its supply of helium from crustal sources. A more likely explanation appears to be that mantle gases are streaming up in that region because an abundant carrier gas is available there, and that in the course of time some pathways have changed, so that pathways of different ages reach different sections of the helium-rich region.

Methane and helium-3

There are many occurrences of methane, both in commercial deposits and in methane seeps, where the associated helium is enriched in helium-3 so as to be identified as primordial and coming from the mantle. The great rift in the Pacific Ocean, called the East Pacific Rise, has been found to emit methane along much of its length, with very hot water (Kim *et al.*, 1983), together with decidedly high ratios of helium-3 to helium-4. Expressing these in terms of the atmospheric ratio, the values there are between 7.5 and 9 times as high. The branch of this rift which goes into the Gulf of California, where both methane and petroleum are present, has a helium ratio of 8 in the same terms. In the rift of the Red Sea, where methane emerges with deep hot water (and where oil in commercial quantities occurs nearby), this ratio of the associated helium is 8.6. The continuation of the same rift into Africa has many lakes that have remarkable occurrences of hydrocarbons. Lake Kivu, which has the world record for the quantity of methane dissolved in its waters, shows a total helium content of roughly a thousand times that which surface water would have brought in from the atmosphere, and the helium-3 proportion is 3 times the atmospheric value (H. Craig, private communication). The lake water has therefore 3000 times the total amount of helium-3 that it could have obtained from the atmosphere. (We shall discuss these rifts again later.)

In Japan there is a rapidly growing commercial gas production from non-sedimentary rock, namely from a porous volcanic rock that is widespread there, called "Green Tuff". This methane has a high helium-3 to helium-4 ratio ranging up to 6 times the atmospheric ratio (Wakita and Sano, 1983). Whatever is argued about the probable origin of all these gases in each of the cases, it is clear that methane and helium-3 are often companions.

Can we identify which gases are commonly the carriers for helium? Since it seems improbable that the helium could habitually have become separated from its carrier gas, we may suspect that the gases usually found in association with it have brought it up. In non-volcanic, helium-rich regions these gases are predominantly nitrogen and methane. The helium/nitrogen ratio is usually higher than the ratio of helium to any other gas. One would judge from this that nitrogen has frequently swept through the longest pathways, and has therefore acquired the highest proportion of helium. The inertness and high thermal stability of nitrogen may well be the reason for this, and it may be the gas that has forced its way up from the hottest, deepest levels in the mantle. A helium-rich region in the outer crust may therefore often be a region which is supplied from a strong source of nitrogen, deeply buried in the mantle. Different pathways from that common source may then result in different helium-3 to helium-4 abundance ratios.

The correlation between helium and methane is not as strong as that between helium and nitrogen, and some precautions have to be taken in evaluating the data. In many cases methane and nitrogen are the two main gases present in a well or a seep, and in that case the higher correlation of helium with nitrogen than with methane would mean that the highest helium proportion would be found in locations with the highest nitrogen, and therefore in these cases with the lowest methane. This will therefore give rise to an apparent anticorrelation of helium and methane. This anticorrelation would occur only in the abundance ratio, but there is no suggestion (and in fact every indication to the contrary) that it would occur in the total quantities transported.

While the gas ratios in any one instance can readily be measured, the total quantities of gas are more difficult to establish. What we can firmly say, though, is that all large commercially significant quantities of helium are in fields of natural gas. If the helium had come from deep, but had merely been caught under a common cap with a gas that had a shallow, local derivation, then we would expect this local gas to have acted merely as a diluent, and the periphery of the gas-producing region to show generally higher proportions of helium. This is not found. In fact, even though we have the strong correlation with nitrogen, very rarely do we find high helium concentrations where the nitrogen is not associated with methane or petroleum. "All the largest helium deposits are related to combustible gases and petroleum" is the summing-up of Nikonov (1973), who has extensively surveyed the global relationships. It has also been noted that the abundances of the other light hydrocarbons, namely ethane, propane and the next few in the series, show a characteristic relationship to the helium proportion in the fields of a given area. It seems very improbable that the helium and the methane can have entered each reservoir from independent sources and established any such relationship. In any case an origin of the helium from the radioactivity in the sediments can be ruled out in many helium-rich regions, just on quantitative grounds. The same is true in many cases also of the radiogenic argon (Pierce *et al.*, 1964). In some cases a shallow origin of the helium can also be ruled out on grounds of the helium-3 proportion.

The relationship of methane in the ground to helium, and even helium escape to the surface, is often so clear that a surface investigation of helium is used commercially as a means of searching for underlying gas and oil fields (Pogorski and Quirt, 1979; Roberts, 1979). If helium and methane had merely found a common reservoir in some instances, this method could not be expected to work. It should be just as likely then to find high helium seepage outside the methane-rich region.

A more probable explanation to account for these observations is that in a gas-prone region, a mix of methane and helium in a certain proportion is emerging from deep below in a system of cracks and fissures, to be arrested in some instances by a sensibly impermeable cap. If this cap can hold much less gas than the quantity coming up from below, then it will spill over the edge, and the ascending gases may then reach the surface in a higher-than-average concentration, in a halo pattern around the cap. This is what is seen in many cases, both for helium and for methane, as well as for various chemical products of seeping methane, which we shall discuss later. An excess of helium seepage over or around a biologically-produced deposit would have no rational explanation. It is quite another matter if both gases derive from deep levels and have ascended by common pathways through the crust.

A particularly clear relationship of helium with methane was recognized in a survey of water wells and springs in Saskatchewan (W. Dyck and C. E. Dunn, 1986). Many sharply defined regions show an enhancement of both methane and helium in water wells. These authors write: "The coincidence of methane and helium anomalies with known tectonic features also indicates fracture leakage from depth, and the possible existence of oil and gas fields". In this case the observations contain no selection in favour of gas fields or caprocks. They represent simply the leakage through the surface, and no explanation for the methane-helium relationship can be offered if the two gases had not come up together from depth.

In those petroliferous regions in which helium shows a decidedly stronger correlation with nitrogen than with methane, it is clear that the bulk of the nitrogen cannot derive from organic sediments. If methane had derived from local organic sediments, then surely it would have as good a chance of picking up helium as does nitrogen, and we could not expect to see so commonly much higher helium values in nitrogen-rich methane wells than in nitrogen-poor ones. A deeper origin for the nitrogen than for the methane is implied in such cases.

The carbon-nitrogen ratio in organic materials has generally a value around 4 or 5, while that ratio in carbonaceous sediments generally appears to be very much larger. It has been argued that the nitrogen probably escapes rather readily and at an early stage of the chemical processing of that buried organic material. Even so, we may wonder whether we have seen enough nitrogen to go with the immense deposits of unoxidized carbon in the rocks, if all this was of biological origin. The helium association now diminishes further, and in some areas to a very low value, the quantity of nitrogen that can be considered as derived from organic sediments.

Regional patterns of mixing ratios

A deep source for gases can also be inferred if the mixing ratios show certain consistencies in gas fields over a large area. If one of the components is helium, and therefore not of biological origin, it will be very difficult to understand that it could establish a well-defined proportion under each one of very many caps in a large area. Yet, such relationships can be found, and in some instances they are remarkably precise. In the Hugoton-Panhandle fields of Kansas and Texas, a major region of commercial production, stretching over a distance of 200 miles, a very clear relationship is seen in the mixing ratios of methane, nitrogen and helium. Data from 1038 wells are available from the U.S. Bureau of Mines, and we have plotted these in figures 9(a) to (i).

The proportions of helium and nitrogen in the mixture are plotted, the third and most abundant gas being methane in all cases. The plot of the 800 wells in Texas shows clearly a narrow range of the ratio of helium to nitrogen (given by the gradient of the straight line to which this plot approximates), and a very sharply defined upper value for this ratio. We must conclude that the entire region has been supplied from a well-mixed common source of nitrogen and helium, in the proportion of 7.5 percent helium and 92.5 percent

Figs 9 a, b, c, d show the systematic relationships in the proportions of the three gases, helium, nitrogen and methane, in the rich gas-fields known as the Hugoton-Panhandle area, stretching from West Texas into Kansas.

Each of the plots shows the proportion of helium and of nitrogen in the commercial wells of the area (1038 wells are represented). Methane, generally the most abundant gas, makes up nearly all the remainder (other hydrocarbons, CO_2, argon etc. making up generally less than 2 per cent, so that the proportion of methane is given approximately by 100% minus the percentage of helium plus nitrogen).

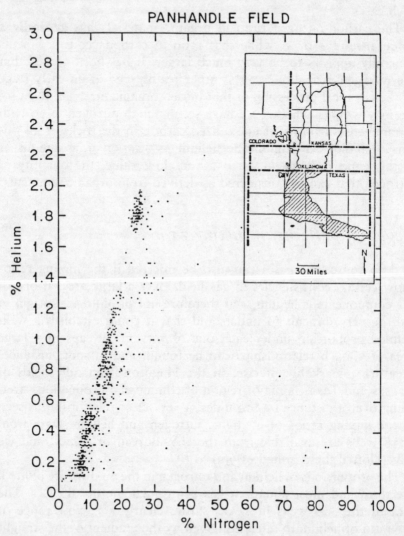

Fig. 9a shows the relationship in the Texas section of the fields. The points fall close to a straight line, implying a nearly constant ratio of helium to nitrogen in all these wells (approximately 0.077), while at the same time showing a large range in the proportion of nitrogen to the remainder (i.e. chiefly methane) from approximately one twentieth to one third.

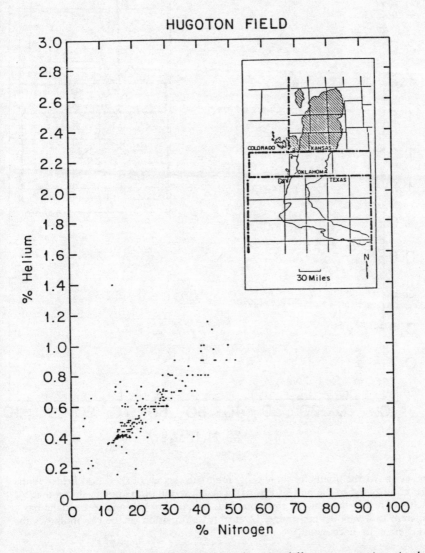

HUGOTON FIELD

Fig. 9b shows a similar relationship but with quite different proportions in the Kansas section of the fields. Here the proportion of helium to nitrogen is approximately 0.021, and again a large range in the proportion of nitrogen to methane is seen.

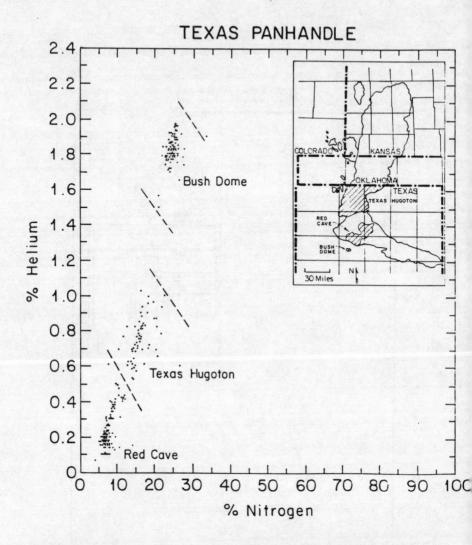

Fig. 9c shows the points for three geographic sub-regions of the Texas fields (points that contributed to the plot of Fig. 9a). The clustering of the points shows that for each sub-region the ratio of the (common) helium-nitrogen mix to methane has a different and well-defined range. (A further sub-division of the region shows the same effect in more detail.)

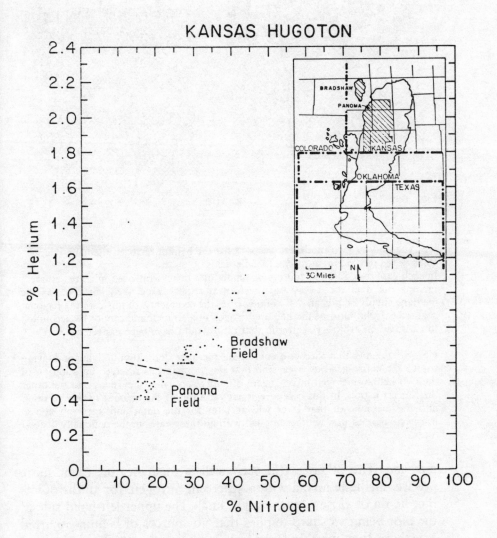

KANSAS HUGOTON

Fig. 9d shows a similar sub-division into two areas of the Kansas section, and again the helium-nitrogen mix characteristic for the Kansas section shows a regionally defined mixing ratio with methane.

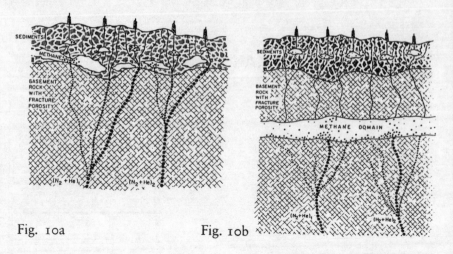

Fig. 10a   Fig. 10b

Figs 10a and b: Two models to account for the helium-nitrogen-methane relation-ships noted in the Hugoton-Pandhandle fields.
In each case two different sources of supply of a helium-nitrogen mix are assumed, fanning out over the Texas and the Kansas region respectively. In Fig. 10a the methane supply is indicated as coming from the sediments. In this case the regional character of the ratio of the helium-nitrogen mix to methane cannot be accounted for; instead an arbitrary scatter in that ratio would have been expected.

Fig. 10b assumes that there is a continuous methane domain underlying and giving rise to the entire gas-rich area; and that this domain has received an addition of the two helium-nitrogen mixes, again one mix supplying the Texas and the other the Kansas region. In this case a regional variation of the amount of this addition into the methane of that layer will account for the data, and any trap above, irrespective of its size, will be supplied with all three gases in the regionally defined ratio.

nitrogen. The other gas in each well is methane (or, to be more precise, methane mixed with 0.3 percent nitrogen, for all the points that lie on or close to the straight line). The upper left-hand side of the plot being so sharp implies that no sources of helium occurred anywhere that were independent of the nitrogen. The less sharp lower side of the distribution may be due either to a selective helium loss (which may well be expected on account of its higher diffusion rate), or to the addition of some smaller extra amounts of nitrogen, either from sources in the ground or from atmospheric contami-nation in the sampling.

We cannot envision a well-mixed reservoir of helium and nitrogen in the sediments that gave a supply to every well of this area. The distribution of radioactively produced helium would be variable and quite unrelated to the distribution of organic material that might have given rise to nitrogen. Only a source of well-mixed helium-nitrogen gas at such a depth that it would infuse the entire region can be held responsible.

But could it be thought that while the helium-nitrogen mix had entered the entire region from deep below, the methane was derived in the sediments? Such an explanation would give no reason why the mixing ratio of the helium-nitrogen gas to the methane should not have a quite arbitrary value for each field.

Gases ascending from deep below come out preferentially through faults or lines of weakness in the basement rocks, and very large variations in their concentration would be expected, and are indeed seen, even over small horizontal distances. The amount of the helium-nitrogen mix that would be expected under any one cap would have to reflect such extreme local variability. The amount of methane derived in the local sediments would be quite unrelated to the helium-nitrogen supply that had happened to emerge there, and therefore the final methane-nitrogen ratio should show a wide and seemingly arbitrary scatter. In fact, the methane-nitrogen ratio is far from arbitrary in the various wells; it shows a clear regional pattern, going systematically from a high value along the southern edge of the fields, to a lower value to the north, and then again to a higher value towards the Oklahoma border. Subdividing the plots of helium against nitrogen for the various areas shows this effect (remembering that the third gas of the mix is predominantly methane). We cannot imagine any way in which a locally-produced gas in the sediments could have entered under each caprock in a well-defined mixing ratio to the deep source gas emerging through cracks from below.

The best explanation for this seems to be that a common helium-nitrogen mix from the deepest level had to traverse a methane-rich layer, whose thickness or methane concentration had a general taper over the region (Fig. 10). The gas penetrating through this, and emerging above, would then have a constant helium-nitrogen ratio, but would have picked up an amount of methane which had a regional dependence. What was caught under each caprock then has the defined mixing ratios without any arbitrariness due to the variable local circumstances. This means of course that only a small proportion of any of the three gases can have an origin in the local sediments.

Other areas show systematic effects of the same nature, but with different values of the ratios. In any one area these ratios seem to follow a common pattern, quite irrespective of the nature and geological age or depth of the formation that has been regarded as the source of the gas, or the present depth of the reservoir; the geographical location seems to be what matters most. Thus the extension of the gas-rich region to the north into Kansas takes us to a domain where again all wells have closely-defined mixing ratios of the same three gases, but here the ratios are different. The supply to each well here comes from one mixed source of methane with 0.18 percent helium in it, in turn mixed with another gas consisting of nitrogen with 1.77 percent helium. (These figures are not individually precise for each well, but are statistically closely defined.) These two individually mixed gases supplied each well in variable, but by no means arbitrary, proportions. Again there is a limited range and a regional dependence of this ratio which requires a similar explanation to the one offered for the Texas fields.

It is interesting to note that all these Kansas fields are very shallow, mostly less than 3000 feet deep, and the crystalline basement is mostly at 3000 feet. Some wells are actually producing from fractured basement rock. Neither of the two component gases could have been supplied over the area from a single source, let alone could two separate mixes have come from two separate shallow reservoirs and then have fed each well with a closely-defined ratio. A sedimentary origin for the methane cannot account for the mixing ratios observed in all these large commercial fields.

As a methane domain, the Hugoton-Panhandle fields, stretching from Kansas through the Oklahoma Panhandle into Texas, form clearly one single connected region of particularly high productivity. The helium-nitrogen mix, however, seems to come from two distinct sources. No helium-nitrogen mix from either of the sources seems to have filled any traps without methane, either inside the productive region or outside it; we do not see any well where nitrogen with 7.5% helium is the principal gas. (Some high nitrogen wells exist, but those lack the helium.) It appears therefore that the source region of the methane is continuous over the entire area, so that the helium-nitrogen gas, coming from much deeper levels, can find no pathway to the surface that avoids the methane region (Fig. 10). The methane, being more abundant than the nitrogen, then becomes the gas that creates the cracks through which the mixture makes its way through the cooler and harder part of the crust. At the deeper levels, below the methane layer, the smaller quantities of the helium-nitrogen gas must suffice to create some mobility.

There are some instances where the role of nitrogen and methane is reversed and where in fact the methane brings up the higher concentration of helium. We would argue in this case that the nitrogen source happens to be at a shallower level in the mantle than that of the methane.

In view of the fact that on an average much more methane than nitrogen is making its way towards the surface, the helium brought up by the methane, though a lower proportion in the stream, is still in total quantity more than all the helium brought up by nitrogen, or any other gas.

Another noble gas, argon, tells a similar story. Like helium, argon is produced mainly by radioactivity in the rocks, in this case by potassium, and it would become available in low concentration only. Like helium, it will be dependent on a carrier fluid for its transportation.

In the atmosphere the argon accumulates. Unlike helium, it is too heavy an atom to evaporate into space, and a concentration of about one percent has built up in the atmosphere. Again, nitrogen seems to be the gas that brings up the highest concentrations, but probably methane the largest total amounts.

Nitrogen emerging from deep levels at the present time, and giving every indication of being primordial, carries up argon at a concentration of between one quarter and one tenth percent. That must be a much higher level than the concentration brought up in the other volatiles, which have amounted to so much more in total quantity. 300 atmospheres of water (or 300 kilogrammes per square centimetre) have been supplied to the surface, and 20 kilogrammes per square centimetre at least of carbon in a fluid have come up. If these volatiles had supplied a similar concentration of argon as the present nitrogen, the atmosphere would contain at least thirty times as much as it does. The conclusion is also here that nitrogen is the most effective sweeping agent, presumably because it has come from the deepest levels, and therefore swept over the longest distances. Methane, like other hydrocarbons, is again in the second place as an argon carrier for concentration, but probably in the first place for the total amount. The budget of primordial water emerging at the present time is still unclear.

While we have argued here about the details of the association between helium, nitrogen and methane, and what they imply, it is just the overall global relationship between helium and methane which establishes the strong presumption that methane is the dominant gas that comes from sufficiently deep levels to have picked up a large amount of helium. If the bulk of the methane had derived from biogenic deposits, there would be absolutely no reason for the observed relationship. The short pathways in sediments that biogenic methane could have maintained could have collected no more helium than pathways of water, which are hundreds of times more common, or pathways of carbon dioxide, which, though less common, are found to be generally much lower in helium. The gases we now find associated with most helium can be presumed to be the ones mainly responsible for bringing it up. On this basis we would judge that a small amount of nitrogen had come up from very deep levels and supplied a high concentration of helium, and that a much larger amount of methane had come from levels that were not quite as deep, but that had come up in much larger quantity and had therefore scavenged on the way up the major quantity of helium that we now find at or near the surface of the Earth.

8

WHERE ARE OIL AND GAS FOUND?

Why the Middle East?

When I first started to investigate the distribution of oil and gas over the globe, I was of course most impressed, like everyone else, by the enormous concentration in the Middle East. Saudi Arabia, the Gulf States, Iran and Iraq dominated the world's energy supplies, and through that, exerted an enormous influence on the world's economy and on global military dispositions. Surely, I thought, this region must by now have been very well investigated and understood, the resources and their derivation well documented, and the conventional explanations must have found strong support there. A theory of the origin of the world's hydrocarbons would surely have to be able to account for the number one hydrocarbon province, before gaining widespread acceptance.

In many discussions and debates, I heard statements to the effect that the area had been enormously favoured by rich biology over long periods; that huge amounts of organic sediments lay buried there; that exceptionally good and dense caprocks held down the rich oil and gas deposits; that in fact all the circumstances necessary for the generation and containment of petroleum had come together in the whole of this area, to create the outstanding wealth of deposits. I was also told that the biological origin of all that petroleum was firmly established by chemical and isotopic studies.

Slowly I came to understand that all these beliefs had arisen in reverse. The presence of all the large petroleum deposits had merely suggested that the province *must have been* favoured in all these ways – if one believed in the bio-origin theory. As for actual documentation, there was nothing that indicated these enormously favourable circumstances. The oil was there in great abundance, but not much else.

For 15 years the concession for Kuwait was offered to the leading oil companies in the world; they all declined, despite the prolific oil seeps in the area. As Wallace Pratt (1952) put it, "they knew there

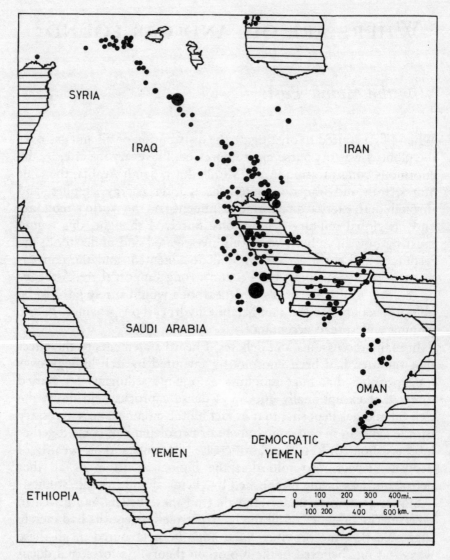

Fig. 11: Mideastern oil and gas fields, stretching from Southeastern Turkey to the opening of the Persian Gulf.
Although it is one connected area that is outstanding in its hydrocarbon supplies, the different parts of it are in quite different geological settings.

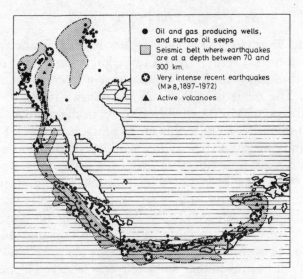

Fig. 12: Relation between volcanoes, earthquakes and petroleum occurrence in Indonesia and Burma. The volcanic and seismic belt is clearly defined from the western tip of New Guinea through the island arc of Java and Sumatra, continuing then to the North into Burma and Southern China. Hydrocarbon production parallels this belt closely through its entire length, despite great differences in the geological settings.

was no oil in Arabia". Then, in 1937, the oil field that was discovered there proved to be by far the world's largest one known at the time.

Not much has changed since then, except that a lot more oil has been discovered. It is now said that 60 percent of the world's recoverable oil is there, concentrated in an area that is only about two percent of the Earth's land surface. But any documentation suggesting that the area was particularly favoured with biological deposits is still missing. What is still more remarkable, is the complete absence of any unifying feature that would explain why all that oil is in one concentrated area of the globe (Fig. 11).

In detail, the oilfields of the area have little in common. Some are in the folded mountains of Iran, some in the flat deposits of the Arabian desert. The oil and the underlying gas fields span over quite different geological epochs, have different reservoir rocks and quite different caprocks. The search for organic source-rocks responsible for the world's largest oilfields has not led to any clear consensus. Sediments of quite different type and age have been suggested here and there, and evidently quite different materials serve as caprocks. The quantities of organic sediments have been regarded as inadequate for the production of all the oil and gas, and would probably be seen to be much more inadequate still if one allowed for the large natural seepage in the area (Barker and Dickey, 1984).

Kent and Warman (1972), who report on their detailed geological studies of the area, write: "It is a remarkable circumstance that the world's richest oil-bearing region is deficient in conventional source rocks. . . . The oil is distributed in reservoirs varying in age from middle Jurassic to Miocene, with maxima in the middle Cretaceous and Oligo-miocene. Despite this range of age and type of reservoir, there is a notable homogeneity in chemical composition of the oils, and there is a presumption that they have a common stratigraphic origin." It has also been noted that "Most of the reservoirs are conspicuously full to the spill level". Coal deposits have also been noted, both in the mountains of Persia and in the flat plains of Arabia, contributing once more to the suggestion that oil and coal have some relationship.

It could not be argued that the whole area is served by particularly good caprocks, since the natural seepage of hydrocarbons to the surface is particularly high there. Already in antiquity many seeps of bitumen and asphalt in the region were known and provided materials needed for building and shipping. The region is not outstanding for good containment and not for good so-called source material, but apparently just for the quantity of hydrocarbons that have entered into every domain.

However detailed some of the explanations may have been for the derivation of the oil in any one of the many fields, these explanations are clearly inadequate, so long as they do not account for the general abundance of the region, or for the common chemistry among all the oils. Surely this cannot just be considered a multiple coincidence.

If the unifying feature of that region cannot be seen in the sediments, then the presumption is that it lies underneath. Presumably the underlying mantle of the Earth in that area happens to be particularly hydrocarbon-rich. The scale of unevenness of mantle composition is recognized in other instances to be very much larger than the scale of crustal features and crustal topography. A hydrocarbon-rich area of the mantle will make the overlying crust oil- and gas-rich by filling and overfilling every available trap, no matter whether these traps are in the flat plains or the steep mountains, whether they are old or young, at deep or at shallow levels. If the sources of petroleum and gas generally lie deep, then one must expect such occurrences, namely that chemically closely related oil deposits may have invaded an area of the crust with no regard to the local features.

One can perhaps go a step further still and speculate that the configuration of the crust in that area is particularly helpful. The large paving stone which is Arabia is remarkably smooth and undistorted, and it is thicker at the western edge where it borders the Red Sea, and thin in the east, at the Tigris Valley and the Persian Gulf. The surface between mantle and crust tends to have the inverse shape of the top surface, since the crust floats on the softer mantle and therefore sinks in deeper, where it is thicker and therefore higher on the surface. The underside of all of Arabia may therefore provide a huge unbroken surface, which guides any fluids coming up from the mantle towards the highest point, where they will then pool and make their way to the surface. The Persian Gulf and Tigris Valley may be just that place. On the other side of the larger plate, the Red Sea and the Sinai Peninsula have also some production and many signs of seepage of hydrocarbons, and the helium analysis has given good reason for ascribing a deep origin, at least to some of the fluids. But on that side, the quantities of hydrocarbons are not so large, because it is not favoured by the guiding effect of the Arabian plate.

This crustal guiding of upwelling fluids is of course only a speculation at this time, but it is worth discussing since it may become possible to confirm it, and since it would then have an influence on the art of prospecting for new oil fields. Seismic measurements of the thickness of the crust are in progress, and they may allow us to map out under which areas the crust has a high spot that may favour the pooling of light ascending fluids.

The Middle East is of course the outstanding example of hydrocarbon occurrence that transcends the local geological settings, but it is certainly not the only one. In fact this seems to be the case for most of the larger of the world's petroleum provinces. In Indonesia, for example, much oil and gas is produced along the island arc stretching from the western tip of New Guinea through Java and Sumatra. Most of the production there parallels the line of the active volcanoes, mostly to the north, which is the inside of the circular arc. Many, or perhaps most, of these oil wells also penetrated coal seams, and it was the local lore among the drillers there that if you went through coal you were sure to hit oil. The line of these producing oil fields, as well as the line of the active volcanoes, follows closely the pattern of the very intense local seismic activity. The active volcanoes tend to be on the arc where the earthquakes are at a depth of between 80 and 120 kilometres. Most of the oilwells are to the inside of the arc, where, in the usual pattern of island arcs, the earthquakes are a little deeper (Fig. 12).

The pattern of this arc, defined by both earthquakes with their well-monitored depth range, and active volcanoes, continues in the west and north, from Sumatra through the Andaman Islands, into Burma and Southern China. There can be no question that it is the same arc, defined by some deep underlying tectonic feature. But the surface topography and geological setting are quite different; from the volcanic islands we have now moved into an area of steeply folded mountains. Yet the oil of Burma follows that same arc pattern into the folded mountains, up to and beyond the Chinese border. Again the unifying feature of this oil and gas province cannot be seen in the crustal features, but in this case it is clearly indicated by volcanoes and earthquakes that it exists at a lower level.

Many petroleum provinces were largely discovered by working along a major tectonic feature, from the location of the first strike. When the gas field of Gröningen in Holland was discovered, interest arose in following the North Sea trench, which stretches from there along the coast of Norway, and up into the far north. The early predictions were that the North Sea was a hopeless location for petroleum, and it is said that one adviser told the British government that he would drink every cup of oil that was obtained out of the North Sea. In fact the search for oil and gas along this tectonic feature proved most fruitful, and it is still being continued and may turn out to be successful up as far as Svalbard (Spitsbergen), or even beyond. Svalbard, incidently, has enormous deposits of coal, but also every indication of large quantities of methane.

Many other large-scale patterns of petroleum occurrence can be mentioned. We have already discussed, in relation to the helium emission, some of the great rift systems. The East Pacific Rise, containing the major crack that runs through the Pacific Ocean, is perhaps the best-documented example of methane emerging from depth in a nonsedimentary region, together with helium high in helium-3, and also carbon dioxide in large abundance. The large heat flow in the area suggests that this is essentially a volcanic domain in which liquid rock is available at comparatively shallow depths. We think it likely that here, as in other volcanic regions, the carbon dioxide is the oxidation product of methane.

The eastern branch of the Pacific rift becomes the Gulf of California and the Imperial Valley. It is there that sediments exist and oil is found in them. One could well regard this as supporting the viewpoint that hydrocarbon fluids emerging from great depth tend to result in the deposition of oil when they move through sediments.

The great rift of the Red Sea and the string of lakes of East Africa present a similar picture. Again, in the Red Sea ocean floor rift, methane and helium enriched in helium-3 are observed. There is oil production on the shores of the Red Sea, and some oil has been recovered along the shores of Lake Albert. Lake Tanganyika and Lake Malawi have elevated levels of methane in their deep waters and appear to be the objects of prospecting activity at the present time. However, the outstanding spot is Lake Kivu, with its high concentration of methane, carbon dioxide and helium enriched in helium-3.

The strange case of Lake Kivu

Lake Kivu is one of the lakes right in the middle of the great rift, where apparently the continent is splitting apart at the present time. It is a lake of about 24,000 square kilometres area, and approximately 500 metres deep. A major faultline, quite possibly the principle crack pulling apart at this time, is known to be running right through it.

The deep waters of this lake have very high levels of methane and carbon dioxide dissolved in them, although neither gas reaches saturation level at the present time (Deuser *et al.*, 1973; Degens *et al.*, 1973). The total quantity of methane in the deep water is estimated at 2 trillion cubic feet (for comparison, the U.S. annual consumption of natural gas is 17 trillion cubic feet). With this it is the world's outstanding body of water for its *total* methane content, and it is also outstanding for the *concentration* of methane in any major body of water.

When the content of helium was investigated, it also turned out to be remarkable. The total amount of helium in the lake water was found to be about 1000 times greater than it would be in ordinary surface water, which, like most other lakes, had equilibrated its helium concentration with the atmosphere. Not only was the total helium so abnormally high, but it was particularly rich in helium-3, the isotope that indicates a deep origin for the helium. In fact the total amount of helium-3 in the lake is some 3000 times greater than it would be in ordinary surface waters. As we have discussed previously, one must suppose that the helium was carried up by some other gases, and one would therefore be inclined to regard methane and carbon dioxide as those carrier gases.

From all this we would be tempted to judge that the lake was extraordinary in all these respects, just because of the single fact that it had an extraordinarily good supply of gases from deep below as a result of the rift. However, there is a snag with this explanation.

The well-known technique of carbon-14 dating can be applied to the methane of the lake. The technique is generally employed to give the age of a carbon deposit from the time at which as a plant it had derived its carbon from atmospheric carbon dioxide. The atmospheric carbon dioxide, having all the time a certain small proportion of carbon-14, unstable with a half-life of 6,800 years, will supply all plants with that proportion, but then, once the carbon is fixed and does not interchange with the atmosphere, the proportion of carbon-14 will decline in accordance with the laws of radioactivity, and the final proportion found in a sample will continue to diminish with time. Any carbon that is coming up from deep in the Earth would be expected to have no carbon-14 with it at all. One would loosely refer to it as "infinitely old".

Now the dilemma in Lake Kivu is that the methane measured by these techniques is not infinitely old. An age of the order of 20,000 years has been attributed to it (Tietze *et al.*, 1980). On the basis of this it was claimed that the methane must derive from organic debris sinking down from the top, and we have therefore the dilemma that the world's outstanding helium lake has resulted from upwelling gases that enter it from below, but perchance it it also the world's outstanding methane lake, produced by quite unrelated biological processes. On the face of it that would seem a very improbable coincidence.

An explanation that avoids this extreme improbability and may fit all the facts involves hydrogen, as well as methane and helium, entering the lake from below. The rift is in an intensely volcanic region, and many hot springs and fumaroles in the general area emit hydrogen. Hydrogen has also been observed burning in the lava that spills out of the large nearby volcano Nyiragongo. Hydrogen bubbling through the sediments of the lake that sink down in the normal biological processes will readily convert some of the carbon to methane. (It has been observed in one instance that hydrogen bubbling through sediments causes the production of 17 times more methane than the plant material would otherwise produce. It is clear that organic debris is usually very limited by the availability of hydrogen for the amount of methane that can be generated. If much hydrogen is available, the enhancement factor might well be even more than 17.)

In volcanic circumstances, at low pressure and high temperature, methane is quite readily dissociated, and the hydrogen seen in this region, as in other volcanic regions, may well be the result of such dissociation. Then in the sediments the hydrogen may combine with fresh carbon of biological origin and generate much larger quantities of methane than similar sediments would have done in the absence of hydrogen. This young methane, together with the old (carbon-14-free) methane, would then make up the mix that gives the appearance of being 20,000 years old. In this case the lake is outstanding for its supply of deep gases, namely helium, carbon dioxide, methane and hydrogen, but it may have no more than the ordinary precipitation of organic debris.

Lake Baikal in Siberia was also created by a rift. We have no knowledge at this time of the gas content of the deep waters of this lake, but we do know that there is gas and oil production around its shore. Some of the production there is indeed from Precambrian rock.

The Kurile Islands and their continuation, the Kamchatka Peninsula to the north, all have strong active volcanism and prominent hydrocarbon seepage nearby. The magnitude of the seepage in the Kuriles has been estimated by Kravtsov (1975), and he discusses how much methane would have escaped if the seepage had been similar to the present one during the 80 or so million years of the existence of these islands. The answer is 380 trillion cubic metres (or 13,500 trillion cubic feet) of methane, which is vastly more than the content of any known gas-producing area of similar dimensions.

Other island arcs, like that of the Caribbean Islands, show a similar relationship between volcanoes, earthquakes and the occurrence of sizable amounts of oil and gas.

One may speculate whether the other island arcs in the world will also provide a supply of hydrocarbons. The remarkably accurate circular arcs made up by the Marianas or by the South Sandwich Islands show the same patterns as other island arcs, namely an ocean trench to the outside, then a line of active volcanoes as one moves inwards across the arc, a deepening of the sources of earthquakes and a line of hydrocarbon occurrence, mainly paralleling the active volcanoes, but to the inside of the arc.

We have already mentioned the global patterns made up by the lines along which earthquakes are common, and the strong relationship these show to commercial hydrocarbon occurrences. Of course there are plenty of hydrocarbon occurrences outside the seismically active areas also, but nevertheless it is clear that there is a strong positive correlation. Of course other explanations than ours have

been advanced for this relationship. It has been said that the seismically active lines are also those on which sediments accumulate in a
way that causes them to be good traps for hydrocarbons. If seismically active areas had been found to be devoid of hydrocarbons, the
explanation that they escaped too readily there would have been an
obvious one. As it is, the presence of hydrocarbons, because of better
capture, is hardly an acceptable explanation when one finds that
those same regions are also the ones of the greatest natural seepage
rate. It is in fact quite generally true that the locations that contain
the greatest quantity of hydrocarbons are also the locations that
show the most seepage. They must therefore be locations where the
supply has been particularly good, and on the whole not locations
favoured by virtue of a superior caprock only.

Proximity to active volcanoes is also worth noting. There are
several volcanoes on whose flanks oil and gas are being produced
in commercial quantities. The north island of New Zealand, which
is generally very gas-rich and which once had the gigantic eruption
that created Lake Taupo, has gas production on the flanks of its
largest presently active volcano, Mount Egmont. Mount Etna in
Sicily has commercial oil and gas production on its flanks. Again
another explanation has been offered for this relationship, namely
that the volcanic heat helps to cure organic sediments and convert
them into hydrocarbons. But in those cases, as in many others, the
quantities of sediments available are very small and the natural
escape rate of the gases in the neighbourhood is very large. It would
be most surprising if such locations were favourable for the
production and retention of substantial quantities of methane or oil.

We have discussed here the geographical distribution of petroliferous areas, but there is another feature about them that we have
already mentioned. It is the vertical distribution of hydrocarbons in
any such area; the fact that it is quite common to find every level
that the drill passes on the way down to contain oil or gas
(Kudryavtsev's rule). The quantities that may be recoverable at the
different levels may vary greatly, as porosity and permeability vary,
but the entire vertical column has hydrocarbons in its pore spaces.
At first sight one might argue that this would just result from a
prolific supply at the deepest level. But when, as often happens, the
deepest level is in fact the crystalline basement or a very ancient
non-fossil-bearing sediment just overlying the basement, then such
an explanation seems inadequate.

This vertical stacking of hydrocarbon domains right down to

basement is seen in the Anadarko Basin in Oklahoma, in the Hugoton Panhandle fields of Kansas, Oklahoma and Texas, in the San Juan Basin of New Mexico, but equally in the Persian Gulf, where, for example, a large gas field at a depth of 17,000 feet was found to underlie the large oil fields at shallow level in Abu Dabi. If organically rich source rocks were held responsible for the hydrocarbons, then we would have to suppose that these locations acquired them vertically stacked at different times in their geological history. On the other hand, if we suppose that the hydrocarbons emanate from below anyway, then vertical stacking would be expected to be a common feature of the distribution of oil and gas.

In the Hugoton fields of Kansas one can pursue this vertical stacking not only down to the basement, but actually into it. Commercial production by numerous wells is from fractured basement. In that area, as perhaps in many others, it is possible to recognize the relationship of the producing shallower wells to the fault patterns in the basement underneath. Landes (1970), who describes the Kansas region writes: "Oil reservoirs occur in every Paleozoic period and even in the Precambrian itself". "Other producing rocks are the . . . weathered and fissured granitic rocks of Precambrian age, immediately underlying the Paleozoic section." "The discovery of oil in commercial quantities in the Precambrian of the Central Kansas Uplift in 1932 really confounded the textbook writers. The discovery well produced at a rate of 1800 barrels daily from fractured Precambrian quartzite. Subsequently more than 50 producing wells have been completed in the basement rock in this area." And again, "The area of greatest promise is the Hugoton Embayment. . . . This Embayment has both a large area and a stratigraphic section that includes potential reservoirs ranging in age from the Cretaceous to and into the basement."

There can be little question that if hydrocarbons are found in an area, then other levels in that same area are worth investigating, irrespective of their age. That includes the crystalline basement, if there is any chance that it may be fractured. It is also worthwhile extending an investigation from a discovery along any deep-seated linear feature that can be identified in the region. Petroleum prospectors have known this for a long time, but it is not clear what explanations they had for it. In our view a deep-seated feature known to allow the upwelling of hydrocarbons in one location is likely to do so in other locations along its length.

9
THE ORIGIN OF PETROLEUM

"We don't care about theories how the oil was made – all we do is to find it and produce it" – that was a statement made to me by a petroleum geologist after he had listened to a discussion about the origin of oil. When then asked whether he would look for oil in the basement rocks, he replied: "Well, of course not, there can't be any source rock in the basement". The theory he had accepted was evidently no longer a theory to him; to discard it and not to require a source rock seemed inconceivable.

Of course the search for oil and gas will always be influenced or even determined by the theory of origin that is adopted. However many wells are drilled, they will still probe only a small fraction of the ground, and of all the levels and formations that the crust of the Earth can offer. Some speculation, guided by theories or beliefs, will always enter into the choice of a site for exploration. Whether it is explicitly understood or not, the theories of the origin of hydrocarbons are affecting the way in which many billions of oil exploration dollars are spent. It is surprising that a subject of such scientific interest and of such immense economic importance has not attracted more research effort, and that the unsolved and problematical aspects are not constantly under investigation and debate.

The last two decades have given us a greatly improved and changed understanding of the chemistry of the other planets and their satellites, of the formation process of the Earth, of the nature of asteroids, comets and meteorites, and it is surely time now to re-examine many beliefs that grew up before this period. As deposits of unoxidized carbon are now recognized to be abundant on other cosmic bodies, the origin of such deposits on the Earth is due for a review. Among those deposits are the deposits of methane, of petroleum, of various forms of coal and carbonaceous shale, and of the ill-defined group of substances called kerogen.

We have seen that outgassing processes from deep levels seem to be at work. We can no longer consider it in any way improbable that fluids are making their way up from below and are constantly enriching the atmosphere and oceans with carbon. The carbon dioxide of the atmosphere and oceans appears to have been replenished many times during the time-span of geology, since otherwise its removal and deposition in the sediments would have depleted it on a timescale of the order of 10,000 years, one millionth part or so of the time-span we see in the geological record.

The huge eruptions of the past that sprayed out diamonds and other deep-seated rocks over the surface, and the explosive volcanic eruptions that spread thousands of cubic kilometres of volcanic ash over continents, all make clear that violent outgassing from deep levels is a continuing process. Nobody can reasonably doubt that. What does seem to be ignored, however, are the gentle forms of outgassing. Thus the gas phenomena associated with earthquakes have generally been discussed as merely due to crustal gases being pushed from one place to another. Volcanoes, and the associated phenomena that occur with them, have been acknowledged to arise from mantle depths, but "cool" outgassing, the streaming up of fluids in the absence of molten rock, has had little or no discussion. Somehow it seems to have been implied that it was impossible. At great depth the rock was thought to be closed as a consequence of the weight of the overburden, and it was therefore thought to be quite impermeable to fluids. The simple notion that at great depth there is a "pressure bath", where fluids and solids coexist at essentially the same pressure, seems to have been largely ignored, even though high concentrations of some minerals occur at deep levels where they can only be ascribed to the transportation of fluids through extensive pore spaces. The formation of diamond, which we have already discussed, is a clear case in point.

Like many other mineral deposits, petroleum can probably be produced in a variety of ways. Nevertheless, it is to be expected that one mode of origin will be the most prevalent one and account for the majority of the deposits.

In our present understanding of Solar System carbon chemistry, it seems easiest to account for the presence of both hydrogen and carbon on the Earth as the products of the hydrocarbons and other carbonaceous molecules that appear to have been plentiful in the Solar System. The next question is the detailed fate that these primary carbonaceous compounds would have suffered after their inclusion in the Earth.

The thermal stability of hydrocarbons

Obviously the stability of the various hydrocarbon molecules in the pressure and temperature regime of the deep Earth is a central issue. High pressure-high temperature experiments are difficult and expensive, and so long as one believed in any case in a biological origin of hydrocarbons, it did not seem very important to know what would happen to those molecules in circumstances which they were not thought to have experienced. The limited experimentation that was carried out showed that methane, instead of decomposing at about 800°C, as it would at 1 atmosphere pressure, was largely stable to 1500°C at 10,000 atmospheres (the pressure in the Earth at a depth of 30 kilometres).

The actual temperature of 30 kilometres depth in most areas of the Earth is less than that. Which of the other hydrocarbon molecules would survive 1500°C at 10 kilobars is not yet known experimentally, let alone the question of the stability and the circumstances at several hundred kilometres depth. Theoretical work exists, and, as we shall see, is suggestive of some degree of stability down to 300 kilometres or so.

There are several separate issues involved. First, which types of molecule are stable at such depth, and at all intermediate values of temperature and pressure, if they are to be able to come up? Secondly, if some are not stable, would they be decomposed into carbon and hydrogen, or would they be converted into other species of hydrocarbon molecules? Quite possibly some of each might occur. These questions will get a definite answer only when a substantial amount of experimental work at high pressures and high temperatures has been carried out.

In our laboratories at Cornell University, and with financial support from the U.S. Gas Research Institute, some of this experimentation has been begun. Changes in the distribution of the molecular species can be demonstrated already in such mild circumstances as a temperature range between 150 and 250°C and a pressure of 1000 atmospheres. The effects there have been in the nature of attaching carbon atoms derived from methane onto different types of heavier petroleum molecules (Gold *et al.*, 1986) It seems certain that higher temperatures and higher pressures will show a great range of such changes and that in practice the distribution of molecular species that might come up from below would suffer many modifications from its original source material.

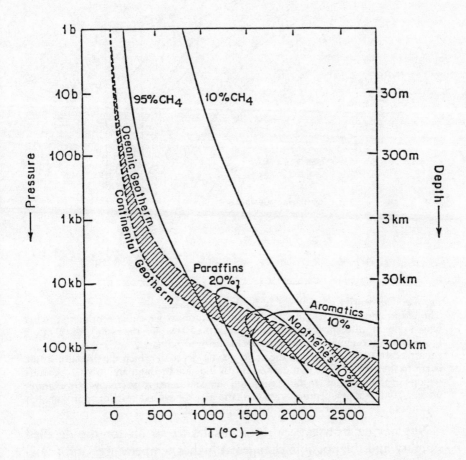

Fig. 13: The stability of hydrocarbons at the temperatures and pressures in the Earth (from Chekaliuk, 1976).

The pressure-temperature regime of the Earth is indicated by the shaded region. Thermodynamic calculations indicate the domain in which various hydrocarbon molecules are stable. According to these calculations, most of the petroleum components would be present in equilibrium at a depth between 100 and 300 km, and methane streaming up could bring a significant fraction towards the surface.

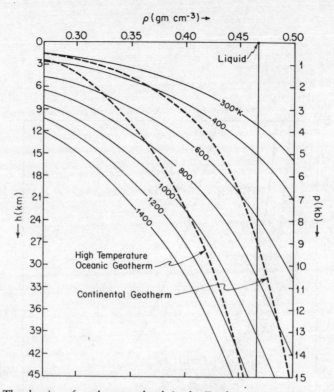

Fig. 14: The density of methane at depth in the Earth.
The graph shows the density of methane (in grams per cubic centimetre; water equals one) at different depths in the crust of the Earth. On the right is the pressure in kilobars (1kb equals 1000 atmospheres or approximately one metric tonne per square centimetre), on the left the approximate depth at which the pressure would occur in the ground. The two dashed curves indicate the high and low temperature regions of the crust under the oceans and the continents respectively. For example under a continent at a depth of 9km methane would have a density of approximately 0.4 times that of water or about half of that of oil.

Not having firm experimental results to go on for the detailed stability and the possible changes at higher temperatures and pressures, we have to do the next best thing and look at the calculations and estimates that are made theoretically. Fig. 13 shows the major conclusions reached by the Soviet thermodynamicist E. B. Chekaliuk (1976), who seems to lead in the study of this problem. His conclusions are that a mix of most of the prominent molecules of petroleum (paraffins, napthenes and aromatics) would be present in equilibrium in a mix of hydrogen and carbon at temperatures of about 1500°C and pressures of 30 kilobars or more. On his graph of stability domains we have plotted the approximate pressure and temperature regime of the nonvolcanic areas of the Earth. His calculations would place the region of the equilibrium production of petroleum molecules in the depth range between 100 and 300 kilometres.

It is perhaps not very important for our present discussions whether these calculations are precisely correct. In the circumstances of high pressure and high temperature, none of the individual molecules would have a permanent stability. There would be a rapid interchange between them, and the final calculations or experiments could merely show the proportion of the different types that exist at any one instant. In detail this distribution will of course depend on the proportion of hydrogen to carbon that is present, on the precise values of temperature and pressure, and on the surfaces of solids that have catalytic effects.

But the mix down there will not be the one that we finally see in the deposits of the crust. During the ascent, as temperature and pressure drop, further changes among the distribution of molecules will take place. Possibly some of the molecules will be split up and cause immobile carbon to be laid down. Some hydrogen-rich and fluid components will continue the upward journey. At sufficiently low temperatures the distribution of the different molecules will "freeze", meaning that essentially no more chemical changes can take place, and it is that mix that we may expect to be delivered to the various traps in the outer crust, which have arrested the upward flow.

In each area there will be somewhat different circumstances for this final "freezing" process, and this will give petroleum and other carbon deposits their regional character. There will also be different concentrations of contaminants such as sulphur, other gases, and various metals, on the long pathways which these fluids have travelled, and this will give them the detailed composition which often shows a regional character of large scale, unrelated to the chemical details of the crust.

There are several reasons for considering methane to be the principal hydrocarbon fluid that enters the crust from below. Firstly it is the most stable of the hydrocarbons and therefore best suited to survive the ascent. It is the most hydrogen-rich, and any loss and deposition of solid carbon on the way up will cause the remainder to become enriched in hydrogen. It is the most fluid of the hydrocarbons, and also the lightest, and for both these reasons it will have the fastest upward migration. Indeed in the deposits of the crust we find large amounts of methane, probably much larger quantities than oil, even though the natural escape rate of methane must be orders of magnitude higher than that of the much more viscous oil. For all these reasons it is probable that methane is the chief fluid that brings with it a variety of the other hydrocarbon molecules in solution.

Even though methane is technically a gas in all the circumstances under discussion, it will behave chemically like a liquid at the high pressures. Already at a depth of 30 kilometres or so, methane has the density that liquid methane would have at low pressures and low temperatures (Fig. 14). High pressure methane will be a good organic solvent, and all the liquid hydrocarbons and many of the solid ones are soluble in it. The solution, if the proportion of methane is high, will be a low-viscosity liquid that can penetrate well through fractures in the rocks.

"Supercritical gases", as high-pressure gases at liquid densities have been called, are recognized now to be often excellent solvents. They have the interesting property of a solubility effect which is very pressure-dependent. This means that in a flow, the substances which such fluids are carrying in solution will be precipitated abruptly in the location of a pressure drop. (Many industrial applications for supercritical gases have been found in recent times, using this convenient property for chemical processing.) In the discussion of the mechanics of outgassing, we have seen that the pressure in pore spaces must develop locations of abrupt pressure changes, and at each such step one may suspect that some of the substances in solution would be shed. All the deeper pressure steps are below the levels that we normally reach by drilling, but we may see some such deposits in rocks that have been uplifted and thereby made accessible. Concentrated deposits of carbon are known in many ancient rocks and may have resulted from this process. The domain which we know well, however, is the one above the last major pressure drop that an upstreaming gas would suffer. It is the domain where gas reaches the "hydrostatic" or "normal" pressure, with which the petroleum industry is familiar. In sediments this last pressure drop for the gases (or first sudden pressure increase for the drillers) is generally between 3 and 6 kilometres depth (9000-18000 ft).

In the conventional discussion of the origin of petroleum, where diffusely distributed organic materials were regarded as the source, there had always remained the difficulty of accounting both for the transport and then for the great concentration of petroleum in the final reservoirs. The usual expectation, after all, would have been that the concentration in the rocks would in general decrease as one goes away from the source material. Yet just the opposite seemed to be seen: an oil reservoir rock might have a porosity of 15 percent entirely filled with oil, while the supposed source rock may have a large total volume, but a concentration of kerogen of only a few percent. In turn, only a few percent of that kerogen would be able to be converted to oil. Concentration factors of several hundred between the source and the reservoir have to be invoked in such an explanation. Why should that occur so commonly?

In the present discussion the high concentration of petroleum in the rocks is accounted for by the flow patterns from deep levels. As we have noted, outgassing can take place only from initially concentrated sources, whose fluid produces a pressure that can fracture the rock. The ascent is therefore channelled to the major cracks that are so forced open, and this gives a lateral concentration. Then, along such pathways, there will be a vertical concentration of the oil shed from the stream of gas, at the pressure discontinuities that such a stream must suffer. In this way the oil may be delivered into a porous region in a sharply concentrated way, and then fill permeable volumes that connect to such a supply point. It is at the last such pressure drop that we find most of the oil, or at shallower levels to which it has subsequently migrated. Since cracks in the crust are often associated with horizontal shear, and this enforces linear patterns, oil-fields are often strung out along lines.

Not only the great concentration of petroleum relative to the source material, but also the fairly sharp cut-off of the occurrence of petroleum below about 15,000 ft depth was a problem for the conventional picture. An attempt was made to account for this cut-off by supposing that the temperature at deeper levels was too high and that oil would not be stable there. On a short timescale oil is stable to very much higher temperatures than the range of 120-180°C, which commonly occurs at 15,000 ft. It was suggested that on the long timescales involved, such low temperatures would already cause instability. Several observations, however, made this explanation appear inadequate. Firstly, gas is often found beneath the oil, but none of the carbonaceous deposits that would have to be there if oil or its supposed organic source material had decomposed to produce all the gas. Secondly, some oil has been found in significant amounts at deeper and much hotter levels. Temperatures as high as 300°C do not seem to have destroyed extractable bitumen in these instances (Price, 1982). It must have existed there for a long time, and if some heavy oil can survive at such a temperature, why not a lot?

Transport by supercritical methane could account not only for the concentration that is deposited above the pressure drop, but also for the fact that often gas, but little other carbonaceous material, is found below. So long as the stream of gas could carry more in solution than it normally contains, it will act as a cleaner and pick up what is there. At a pressure drop it may then shed these substances. In this way the pathways leading to a certain deposit may be particularly free from it, and therefore difficult to trace. A similar effect impedes the tracing of many fluid-deposited minerals also.

The biological markers

How could such a deep origin of petroleum be reconciled with the biological evidence? Sir Robert Robinson wanted a "primordial mix to which bioproducts have been added". In most cases, a very modest addition of bioproducts of one percent or less would suffice. In some cases, however, a much larger addition would seem to be required. The correct theory must be able to account for a great range, from almost complete absence, up to a very prominent presence of biological materials in different petroleums.

There are three effects that have been regarded as specifically biological. First, there is the preference among the chain molecules in petroleum of an odd, rather than an even number of carbon atoms in a chain, an effect which could arise from a certain conversion of biological molecules and which may be difficult to account for without some biological action. Secondly, there is the presence of optical activity – that is, an asymmetry between the number of right-handed and left-handed molecules of the same type, not just in one class of molecules in petroleum, but, where it is present, usually in quite a large number of different types. Thirdly, there are the specific complex organic molecules that are characteristic of biological materials. What is now known about each of these effects?

Let us discuss first the specific biological molecules. We have to be careful to exclude from consideration a large number of complex molecules which occur in biological materials, but which seem to be built up rather readily also in the absence of biological processes. Biology, when it developed, may have adopted some of these molecules for its own purposes. Now, on the biologically contaminated Earth, they would all be regarded as biomolecules.

The molecules in petroleum that have been regarded as the

strongest biological indicators are the porphyrins and the isoprenoids pristane and phytane. Porphyrins are a class of molecules containing carbon, hydrogen, nitrogen and a metal atom, in a well-defined structural relationship. It has been argued that the porphyrins in oil result from chlorophyll of plants, and from the oxygen-carrying dyes in the blood of animals, and there is thought to be a close resemblance in molecular structure. But there are problems with this explanation. Porphyrin-type molecules are found in meteorites and almost certainly are not of biological derivation there (Hodgson and Baker, 1969). Certain types of porphyrins can readily be manufactured in the laboratory from hydrocarbons and nitrogen, and it is not at all clear that we understand which types could be produced in the high-pressure, high-temperature circumstances in the Earth. The porphyrins in petroleum almost invariably have nickel and vanadium as the metal atom, while the biological porphyrins, which are regarded as their predecessors, would have been constructed with magnesium or iron. It is argued that an interchange would have taken place and that nickel and vanadium would generally replace the magnesium and iron. But there are serious doubts that this process would always go to completion. There must have been many circumstances in which the nickel and vanadium were not available in sufficient quantity to make the exchange. Furthermore, why are the oils of some regions enormously rich in nickel and vanadium porphyrins, while in other regions they are not? There clearly remain problems with this interpretation.

The molecules pristane and phytane make a clearer case that biomolecules have entered most petroleums. They represent only a small fraction of the mass and could be merely a contamination picked up as the petroleum soaked through organic materials in the sediments. But there is evidence now of a much more direct contribution of biology to most petroleums: microbial alteration.

While some types of microbial alteration have been identified, and some oils, usually shallow ones, are described as "biodegraded", there is now strong evidence that *most* petroleum has undergone substantial bacterial alteration. Recent investigations (Ourisson *et al.*, 1984), have shown that a group of molecules, called hopanoids, are prominent in all the numerous samples of petroleum and coal which have been tested. These molecules are normally a component of the cell walls of bacteria, and the inference which the investigators make is that all the deposits of petroleum and coal have undergone a substantial treatment by bacteria. One can no longer argue that the presence of specific biomolecules points to a biological source material; instead, it is clear that a substantial biological component was introduced by bacteria, irrespective of the original derivation of the bulk substance. No one supposes that the bacteria were themselves the origin of the bulk material; they did not lay down the deposits of carbonaceous and hydrocarbon substances. What they did, it seems, is to digest much of them, and leave the hopanoid molecules behind, and almost certainly many other specific biomolecules, including the biologically common pristane and phytane.

What about the effects of optical activity, and the odd-even carbon number? It has long been known that these effects are often totally absent, especially in deeper and warmer oils. On the basis of a biological origin of the bulk material, this would be quite hard to explain. In order to eradicate the preference for chain molecules with an odd number of carbon atoms, until this effect is at the percent level where it escapes detection, the average molecule would have to be cut and reassembled several times, so that only a small fraction of the original molecules would survive. Major petroleum-forming chemical processes would have to go on in the sediments, and the final product would be predominantly determined by these processes, and hardly by the detailed structure of the source material.

Equally, the absence of optical activity in some petroleums would require a very thorough chemical destruction if the source material were biological and therefore quite unsymmetrical. It is not just one class of molecules that possess these asymmetries, but in some oils quite a large number of different ones, some causing right-hand rotation and some left-hand. In some other oils the effect is totally absent. Would all the different molecules that are responsible for the effect be destroyed in the same circumstances? Or would some oils of biological origin not have any of these asymmetries, while others (and most biological materials) have many?

What has been found is that there is a sharp transition from oils showing both the odd-even effect and the optical activity in many of its components, to oils showing none of this. Fig. 15 shows the results of one such set of observations (Philippi, 1977). The transition seems to be dictated by temperature: oils whose reservoir temperature was below 66°C are seen to possess optical activity, while it is absent in the same regions in oils with a higher reservoir temperature. 66°C is much too low a temperature to cause a wholesale destruction of many different molecules of petroleum. It is, on the other hand, a temperature that might well be the upper limit for the viability of a certain class of bacteria. It seems for this reason much more probable that the presence of the asymmetries is due to a particular bacterial action which does not take place above 66°C, rather than to think that all petroleum commences with these asymmetries, and that they could be so completely removed by the application of such a modest temperature. In this case the bacterial action seems to be the preferential removal of certain molecules, leaving the remainder with the asymmetries noted.

We could do an experiment: we could pour a barrel of artificially made oil, totally nonbiological, into the ground and leave it for a year or two. Then we could take samples of it and analyse them for all the biological signs: optical activity, bio-markers, odd-even carbon number of the carbon chains; I am sure all the signs would be there, and the oil would just as readily be identified as biologically derived as many a natural petroleum.

At this stage, there is no information to suggest that any other intervention of biology is required to produce all the evidence of biology in petroleum that has been noted. What we seem to have is

Fig. 15: Optical activity in different fractions of petroleum, in different temperature reservoirs (from Philippi, 1977).

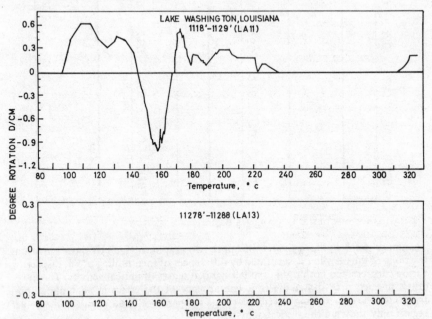

Fig. 15a is for oils in the Lake Washington field in Louisiana.

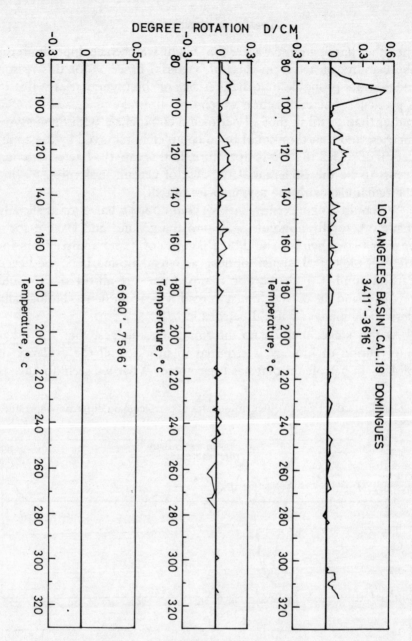

Fig. 15b is for oils in the Los Angeles Basin.

The upper graph in Fig. 15a and the upper two graphs in Fig. 15b show numerous fractions of the petroleum, separated by differences of their boiling point, displaying various degrees and both right- and left-handed senses of optical activity. The lower graph in each case shows the complete absence of the effect in all boiling-point fractions. The sharp change-over occurs at a temperature of 66°C, oils in the cooler region only showing the optical activity.

indeed "a primordial hydrocarbon mixture to which bio-products have been added".

The consequences of methane streaming up through the crust

We can now summarize all the various consequences that can be expected if methane and other hydrocarbon molecules are liberated at some depth in the mantle (like perhaps the depth of diamond formation of 150-300 kilometres) and then make their way up toward the surface through the crust. Such a stream, where it is in solid rock, has to form the pressure domain system, where the fluid pressure is always close to the pressure in the rock and approximates to it in the stepwise pattern we discussed earlier. Where insufficient quantities of fluids are liberated at depth, so that no system of fracture porosity can be created, the fluids will remain down there and we shall not know much about them. Where an upward percolation is taking place, we may expect to be able to identify the consequences of a number of processes.

(1) The methane stream may bring with it in solution a variety of heavier hydrocarbon molecules from the equilibrium distribution that existed at depth. On the way up, temperatures and pressures may be high enough to change the chemical equilibrium between the different types of molecules, until eventually, at shallower levels, the temperature is low enough to "freeze" the distribution. This is then the configuration that we will find in the deposits, except for the effects of later, chiefly biological alteration. The detailed distribution of the molecular species will thus depend on the pressure and temperature circumstances of the ascending pathways, as well as on the minerals that exert a catalytic action. In addition, other fluids, for example helium and other noble gases, may be picked up by this stream. Sulphur, nitrogen and a number of metals that can form organometallic complexes may also join. A variety of substances may thus be brought up that can otherwise be quite sparsely distributed in the mantle or crust. These will be substances that are liquids or gases at mantle temperatures and pressures, or substances that are soluble or form soluble compounds in the high pressure methane.

Such collection and transportation is of course similar to the well-known mineralization processes, where water is the transporting fluid. What we are suggesting is that high pressure methane may equally act as a mineralizing fluid, and concentrations in the sediments not only of petroleum, but of mercury, vanadium, nickel, cobalt, copper, molybdenum, uranium and many other metals that form organometallic compounds, may have been emplaced in this fashion. Patterns of abundance of such components in petroleum have been observed, stretching over regions far larger than those defined by sedimentary basins, or indeed by any characteristic feature of the crust. This can be understood in terms of a large-scale mantle heterogeneity whose pattern bears little relation to the overlying crust.

Regions where the petroleum and natural gas are high in helium clearly show such patterns as we have noted. Mercury, which at mantle temperatures will be a volatile, is also at higher than normal concentrations in some large petroliferous regions (Rudakov, 1973; Antesiforov *et al.*, 1983). Deposits of the mercury ore cinnabar (mercury sulphide) are often found together with deposits of tar, and the concentration of mercury vapour, though quite low, is still orders of magnitude higher in some natural gases than in the surrounding rocks.

Vanadium is particularly enriched in almost all the oils of the whole of South America (Kapo, 1978). In part this is present as the organometallic compound of vanadium porphyrin, but apparently in other, as yet undetermined, organometallic compounds as well. Nickel, another metal that forms organometallic compounds readily, and several other metals show similarly large regional patterns of concentration in petroleum.

(2) Methane and other hydrocarbon molecules may dissociate and provide a source of hydrogen.

If the pathway is too hot or the pressure too low, a partial dissociation may take place which will leave carbon deposited at deep levels as diamond, as we have mentioned, and at shallower levels as graphite, or various coal-like substances. The hydrogen, which is chemically very reactive, may do many things: it may form hydrogen sulphide from any sulphur that is in the rocks; it may of course oxidize to water; or it may form methane again from carbon that is present at shallower levels.

(3) Regions of upstreaming methane will favour the deposition and preservation of carbonaceous materials from biological sources.

While in most cases plant debris is oxidized, with oxygen available from the air or dissolved in groundwater, and the carbon therefore turned back into carbon dioxide and returned to the atmosphere, this process is less likely to be complete where methane emerges from below. Bacteria that derive their energy from the oxidation of such methane tend to deplete the local sources of oxygen, and thereby the oxidation of plant debris is hindered. The plants that grow in such an area may then, when they die, fail to decay completely, and their carbon may accumulate in the deposits they form. This produces one of the interesting relationships between biogenic and abiogenic carbon. Clearly iden ifiable plant debris, like peat or lignite (brown coal in which the structure of the original plants can still be seen), may be found to overlie deposits of oil or gas, giving perhaps the suggestion that plant deposits in earlier times had also been responsible for the deeper carbonaceous materials. When those earlier deposits are several tens of millions of years old, it is not always obvious why just the same location should have been favoured a second time. Local topography and climate, as well as the types of plants, would all have been different. What may have survived over the whole interval of time is the upwelling of a flow of hydrocarbons from a deep source, and it is this that may have supplied the oil and gas, and possibly other carbonaceous deposits at the deeper levels; and it may have maintained the tendency to keep oxygen away and thereby preserve the carbon of any plant debris that happened to be deposited on the surface.

It does not matter much for this purpose whether plants grow fast or slowly in an area. If the ground water has been depleted of oxygen, so that bacteria cannot reoxidize the debris and return the carbon to the atmosphere, then this carbon will accumulate. It will clearly be biological material, but the reason for its accumulation would lie in the circumstances created by the nonbiological hydro-carbons and hydrogen of the area.

There are many locations where one may suspect such a conspiracy between biology and the deep Earth. Large peat deposits of Sumatra are above the oil- and gas-rich regions there. Some lignite deposits, for example a small patch in South America – the north shore of the Straits of Magellan on the Atlantic side – have commer-cial deposits of oil and gas just underneath them. The neighbouring Tierra del Fuego – Land of Fire – may have been named that by Magellan when he saw flames issuing from the ground. This phenomenon has been reported from the area in recent times and is well known in several other parts of the globe, where methane streams out and ignites spontaneously.

The accumulation of peat has often puzzled investigators. It is of course true that it occurs in areas where the ground water is low in oxygen, but it is not clear which is the cause and which the effect. Once there is an accumulation of peat, then bacteria trying to oxidize it will maintain low oxygen levels in that neighbourhood. But why did some areas get into that condition, when other similar areas did not?

The Falkland Islands, with a low rate of growth, have such deposits of peat and they have puzzled investigators (Smith and Clymo, 1984). Similar cases exist in Alaska and northern Canada, with equally low rates of growth of vegetation, but also in tropical regions of very lush vegetation, such as Sumatra. Of course in some cases it may be truly the local formation of a swampland, in which more carbon from plants is deposited than the groundwater can oxidize. But in many other cases we may be seeing the shortage of oxygen caused by the upwelling of gases from below. It is of interest that in some peat fields there is an excess of sulphur beyond that which the plants could have contained, and also that there is a gradation with depth, and therefore with age, of some chemical and isotopic characteristics of the material, which suggests that it has been gradually enriched by some fluids that brought in carbon and sulphur from an external source.

(4) Hydrogen streaming up and originating from the dissociation of methane at a deeper level may help to turn biological debris into hydrocarbons.

The bulk of unoxidized plant debris that is buried has too low a ratio of hydrogen to carbon to be turned into a gaseous or liquid hydrocarbon substance. An external supply of hydrogen can therefore greatly favour such a conversion. (We discussed this in relation to Lake Kivu in the African Rift Valley.)

(5) Methane may be partly dissociated and partly oxidized on the way up, so that carbon monoxide and hydrogen are made available. From these a large range of petroleum molecules can be built up by a well-known process which is used for the industrial production of hydrocarbon molecules (the Fischer-Tropsch process). Petroleum may therefore contain some products derived by a build-up from small molecules, and not only the molecules brought up from the very hot deep layers. A variety of other chemical reactions involving free hydrogen may also occur. If nitrogen is present, a range of compounds of hydrogen, carbon, oxygen and nitrogen are likely to be assembled, many of them similar to molecules that are common

in biological materials. It is the build-up of such molecules in ascending fluids and in the presence of natural catalysts that may have been a vital component in the evolution of life, as suggested by Sylvester-Bradley (1964, 1971).

(6) Bacteria using methane and petroleum as their energy source may leave a variety of biomolecules in the deposits. In addition petroleum may acquire biomolecules by dissolving biological materials that were deposited in the sediments.

Another line of evidence about the origin of the hydrocarbons is obtained from a study of the distribution of the two stable carbon isotopes. This is in itself a very complex subject, and although it provides many interesting clues, it does not lead at this time to any firm conclusion. We shall discuss some aspects of this in the appendix.

There are several other arguments that have been advanced in favour of the conventional biogenic viewpoint. "If oil and gas come up from deep in the Earth, why does one find them chiefly in the sediments?" The answer is that oil and gas are only found where people have looked, and that has been almost exclusively in the sediments. Secondly, sediments are generally more porous than the basement rocks, and therefore they contain much more volume in which these fluids could be stored and kept away from the oxidizing atmosphere. As we have already noted, there are some deposits known in the basement rocks, but such evidence has been ignored or attributed to migration from neighbouring or even from overlying sediments.

How much from above and how much from below?

We have discussed how petroleum and natural gas may be the products of the outgassing process which brings up carbon compounds from deep down in the Earth. We suppose that this process is the one that has supplied all the carbon which is so abundant on the surface and in the sediments, and which amounts to something of the order of 20 kilograms per square centimetre for the global average surface. Of this the atmospheric and oceanic carbon dioxide, and the active biosphere, contain only about 8 grammes per square centimetre. The entire remainder, which is the bulk of the surface carbon, is in long-lived or permanent deposits, either as calcium carbonate (limestone) or as calcium magnesium carbonate (dolomite) in the sediments, as calcite cements (also calcium carbonate), both in sediments and in rocks of the crystalline basement, and in the various forms of unoxidized carbon, such as graphite, coal, the substances referred to as kerogen, and of course petroleum, tar and methane. If all that carbon has originally been supplied from below, which deposits were formed by that upward stream and which by the deposition, through biology or otherwise, of carbon that had reached the atmosphere?

The greatest quantity, about 85 percent of all the deposited carbon, is in the form of the carbonate rocks (Ronov and Yaroshev-skiy, 1976), and they mostly give clear evidence of having been formed by a precipitation in water. They often contain marine fossils, and in fact the carbonate fossils themselves make up a significant fraction of all limestone. There can be no question that all this derived from atmospheric carbon dioxide which intermixes on a short timescale with oceanic carbon dioxide, and with the carbon of all the plants on land and in the ocean. This pool of carbon would be depleted by the deposition of limestone and the other sediments in a time as short as 10,000 years if there were not a constant resupply of fresh carbon from below.

The calcite cements that are found, both in sediments and in basement rocks, have for the most part no such clearly identified origin. They may be produced by ground water carrying atmospheric carbon dioxide down, but there is evidence that they are often produced by fluids that have come up through cracks from deep levels and they are often found in association with petroleum. The carbon isotope values (as discussed in the appendix) of such calcite cements, both in sediments and in the basement, often show that the carbon did not come from the atmospheric pool of carbon dioxide, but from some other source. In petroleum-bearing areas this source is thought to be methane and other hydrocarbons, oxidized by bacteria in the ground to carbon dioxide and then laid down with local calcium oxide as the final calcite (Donovan, *et al.*, 1974). This explanation is strengthened not only by finding a close relationship between such calcite cements and underlying oil and gas fields, but also by finding that this calcite shows a vertical distribution going right through quite different horizontally layered sediments.

Where similar calcite, also with a nonatmospheric isotope signature, is seen in crystalline basement, it is difficult to find an explanation within the conventional theories. If, however, methane is generally ascending from below, there is no reason why the same explanation that fits the facts in the sediments would not also explain these calcite cements in the crystalline rock. Since they are found to be common in large areas of the ancient granites of Canada, Scandinavia and Siberia, this would suggest that some seepage of methane through cracks in the shield areas has been a common phenomenon.

The rocks of the crystalline basement contain also some quantities of carbonaceous materials – unoxidized carbon – usually distributed in finely divided form. The average concentration of this carbon has been quoted as approximately 200 parts per million in the average of all the basement rocks (Hoefs, 1972). With an average thickness of the crustal rocks of 20 kilometres, this would amount to the not insignificant quantity of 1 kilogram per square centimetre, compared with the 20 kilograms of all forms of sedimentary carbon. In the conventional picture, where all unoxidized carbon in the ground is attributed to plant debris, it would be quite hard to see how this material could have penetrated so deeply and uniformly into the crystalline rocks. It is a material which is not in chemical equilibrium with the rock, and if those same rocks were now brought to their melting points and allowed to crystallize again, most of this carbon would end up being oxidized with oxygen available in the rocks. Quite possibly, then, this carbon is also an indication of upwelling methane, either during the time of crystallization of the rocks, or through fine cracks at a later time.

Coal and kerogen

Now, what about the large quantities of coal, and the more diffusely distributed coal-like materials that are called kerogen? These deposits have been regarded as so clearly of biological origin that this assumption was taken as the basis of arguments about other carbonaceous materials. When kerogen was found near a deposit of petroleum, it was regarded as the biological source material that had given off this petroleum. When chemical or isotopic similarities between the kerogen and the neighbouring petroleum were established, this then was regarded as a proof of the biological origin of the petroleum. When all this was found in sediments which contained some fossils – and most sediments do – then the case seemed settled. Coal and carbonaceous shale (a widespread common form of shale whose tiny pore spaces are filled with a tar-like material) were similarly identified without further proof as derived from biological deposits. "You cannot argue with a fossil" was a remark thrown at me at a lecture on the subject. It is true that you cannot argue about the biological nature of the fossil, but you certainly can argue what its presence implies for the surrounding material.

Fossils in coal are usually less common than in the surrounding sedimentary rocks. Within a coal seam they seem to be more common on the ceiling of that seam than in the interior. There is a great variability in the quantity of fossils in coals of the same type, and some coals, as for example the coals of Alaska, are almost free from fossils.

But it is the nature of coal fossils that make it so hard to regard them as the indicator of the biological material that produced the coal. Coal fossils are generally "infusion" or "replacement" fossils, meaning that the structure of an organism has been preserved but the substance has been largely replaced by solids that must have entered that structure as liquids or gases. In some areas the fossils found in coal can be picked out readily because they are made of beautiful golden pyrite (fool's gold). The plant which the fossil represents provided the shape and the space, and sometimes even remarkably fine details of the structure, but it was iron sulphide that came to fill all the spaces the plant once occupied.

Another class of coal fossils are solidly filled with coal. Again, we can be sure that the plant did not grow out of a material which was, like the present coal, composed of 90 percent carbon. It grew out of just a few percent carbon and the rest of the mass and volume was oxygen, hydrogen, nitrogen and other elements. The present fossil, however, may often be structurally remarkably perfect, hardly compressed and with many fine details still clearly visible under the microscope, and yet that structure is filled with the same concentration of coal, as far as one can tell, as the surrounding nonfossiliferous material of the entire coal seam. To think that this fossil is an indicator of the type of plant that laid down the homogeneous coal of the seam leads us to two dilemmas. Firstly, why did the odd fossil retain its structure with perfection, when other, presumably much larger, quantities of such debris adjoining it were so completely destroyed that no structure can be identified at all? Would it not be strange for one leaf or one twig to have its shape perfectly preserved and for all other leaves and twigs in the same assemblage to be turned into a uniform mass?

Secondly there is the other dilemma: how did the coaly material enter the structure of the fossil without destroying it? As solid coal? It certainly could not do that. Like the pyrite fossil, the coal fossil is an infusion product, and the coaly material must have been at some stage in the form of a sufficiently thin liquid so that it could penetrate into the structure of the plant debris. Just as we recognize petrified wood as having been petrified by an infusion process, so coalified wood has similarly been infused.

But if the substance now found inside the structure of the fossil is the same as the homogeneous coal exterior to it, then we can readily suppose that all that homogeneous coal was also at one time in the form of a fluid – liquid or gas – before it laid down the seam. Is that indeed a possibility? Is it possible that the coal seam, like the coal fossil within it, is the product of a deposition from a fluid flow?

There are many other problems that have puzzled the investigators of coal. If coal was laid down from the debris of plants that grew in swamps, then for the quantity of carbon in some coal seams one would have required enormously deep swamps to have developed with an extremely low content of insoluble minerals. Some coal seams are as much as 100 feet thick, and the mineral content may be as low as 4 percent. The bulk of the material is just carbon, with a little hydrogen, oxygen and sulphur mixed in various compounds. For a swamp to produce such a seam, it would need to have grown to a depth of 1000 feet, with a mineral content in that volume of less than 1 percent. No such swamps are recognized, and it seems unlikely that they could ever be created or that plants would grow in such circumstances. The ratio of minerals to carbon in any present-day accumulation of plant debris is a very much higher one, and accumulations of the quantities of biomass carbon necessary to account for major coal seams are not found anywhere.

It is by no means easy to understand the origin of coal, and it is certainly not a subject to be brushed off as fully explained by biological deposits at this stage. With this degree of uncertainty one cannot claim that the deposition of the huge amounts of coal we now find is itself a proof of the previous existence of gigantic swamp-lands, of a magnitude which we no longer have, nor can the frequent occurrence of a succession of coal seams, one above the other, be held to prove the frequent occurrence of an alternation in water levels over long stretches of geological time. These, and many other interpretations that have come to be accepted, are still all dependent on the correct interpretation of the origin of coal, and perhaps of most of the carbonaceous deposits.

Could a stream of liquid or gaseous hydrocarbons coming up in an area have laid down a number of coal seams, or could it have enriched with carbon a number of seams of plant debris of more normal size and mineral content?

There is a chemical property of carbon which is very relevant in this context. A deposit of solid carbon, such as soot, for example, acts as a catalyst for the further deposition of soot from methane or other hydrocarbons. It is a catalyst for the dissociation of the hydrocarbons, and where other circumstances, such as temperature and pressure, would come near to causing the dissociation and the deposition of carbon, the presence of some carbon will initiate it. This means that in an area of upstreaming hydrocarbons there will be a great tendency for carbon deposits to grow to great concentrations, since their very presence is instrumental in laying down more of the same stuff. Perhaps perfectly "ordinary" biological deposits in a sedimentary layer, with normal admixtures of minerals, would act as a starter. Some of the upstreaming methane would be dissociated there, the fossils would be filled with carbon, and as more carbon accumulates, it would serve to cause the further dissociation and further accumulation of carbon. In the end the quantities of carbon from plant material may be an insignificant fraction, and the carbon-to-mineral ratio may have reached values that never occur in surface vegetation. The vertical stacking of coal seams then merely attests to the area being one in which methane outgassing has been going on over extended periods, and in which the circumstances have been mildly in favour of dissociation and carbon deposition.

The detailed properties of different coal deposits may of course still depend on temperature and pressure at the time of formation and the detailed chemistry of the local materials and of the upwelling hydrocarbons, and on the microbial alterations that occurred at the time of deposition or subsequently. Indeed, microbial action might be of major significance in the dissociation process and the observations that coal, like petroleum, generally contain the hopanoid molecules (Ourisson, *et al.*, 1984) would suggest this.

It has often been noted that some coal fields contain and produce more methane than could possibly be produced by the existing coal. It is true of course that coal could give off methane, but one could not expect that more than a small fraction of the hydrogen content of the coal could ever be assembled into methane molecules, and this places a severe limitation on the maximum total methane production that is possible.

Methane will leak out of coal seams into the surrounding rock and out towards the surface, unless there happens to be a particularly tight caprock in the way. A characteristic diffusion time of methane to get out of a coal seam of 6 feet thickness into the surrounding rock is a time of the order of 10,000 years. The amount of methane released by the mining operation should then be no more than the amount that would have seeped out if left to itself in 10,000 years. If a coal deposit is 50 million years old, a mining operation should not release more than 1 part in 5,000 (i.e. 10,000/50 million years) of the amount that would have been produced by the coal and lost over its lifetime. On that basis it should be very rare for any excavated volumes in a coal mine ever to reach combustible levels of methane. Yet even with a very fast enforced air flow, too many mines have methane explosions. Coal mining on Hokkaido, Japan's north island, has come to a standstill because the world's best ventilated coal mines there could not avoid major explosions.

These considerations seem to indicate that methane has entered many coal seams after their formation, or is entering them at the present time.

The methane excess problems would be aggravated if one took the conventional picture, where buried plant debris had to be heated in order to be turned into coal. In that case, at the high temperature, the methane production and its diffusion would both be greatly accelerated, and the amount of methane that could be generated later in that coal at a low temperature would be much smaller.

The geographical distribution of coal deposits poses another problem for the conventional theories. It is assumed that oil and coal are the result of completely different types of deposits laid down in quite different circumstances and, in many regions where both occur, at quite different times. No detailed relationship in the geographical distribution of the two substances would thus be expected. In fact Arthur Holmes, in his famous textbook (1944-1955), writes: ". . . no significant lateral connection between coal seams and oil pools has anywhere been traced. The two may occur in close association by some accident of faulting, and one may lie above the other in a sequence of varied strata, but in neither case has the association any bearing on the origin of oil."

But in fact, as the oil and coal maps of the world have become more complete, it has become clear that there is a close relationship.

The coal and oil maps of South America are quite striking in this respect. Indonesia is another striking example. The local lore among those who drilled for oil there was, "Once we hit coal, we knew we were going to hit oil". In Wyoming, some coal is actually within the oil reservoirs, and in many sedimentary basins, including the San Juan Basin of New Mexico and the Anadarko Basin of Oklahoma, coal overlies oil and gas. Alaska, Iran, Saudi Arabia, the Ural Mountains, all known for their oil fields, also possess large amounts of coal. The same is true of many other major oil-producing areas, such as Venezuela and neighbouring Colombia, Pennsylvania and the Appalachian Mountains, and so on.

The theory that coal was derived by the heating of plant debris runs into problems also when other features of the locality give independent indications of the peak temperatures that have been reached. In many cases oil and gas occur deep underneath the coal in the same area, and it seems highly improbable that the temperatures for coalification of plant material could have been reached higher up, and yet much lower down the temperature would not have exceeded that at which oil would have been destroyed or gas allowed to escape. Yet in many of the areas we have just mentioned, this is the relationship we see. If this occurred just in isolated instances, one might invoke some improbable condition as having occurred there; but in fact it seems to be the rule rather than the exception.

Some of the fossils in the coal similarly give indications that temperature was not responsible for the coalification process. It is not uncommon to find lumps of carbonate rock within a coal seam and, on breaking them open, to find fossils containing wood, not black but light in colour, and showing no signs of a coalification process. Similarly, it is reported that in the coal of the Donetz Basin of the Ukraine there are some fossilized tree trunks that span through a coal seam from the carbonate rock above to that below, and those fossils are coalified where they are within the coal seam, but are not coalified where they are in the carbonate. In all these cases the temperature that the fossils suffered must have been the same as that of the coal. Heating a wood fossil in just the same manner as the local coal did not turn it into coal. A circumstance other than heat must be responsible, and it must be a process which can be prevented by the presence of another type of rock. The dissociation of fluid hydrocarbons might be such a process.

The trace element content of some coals also has some bearing on the question of origin. Like petroleum, coal also has gathered up within itself or its immediate surroundings sometimes very substantial concentrations of elements that are otherwise quite rare. Germanium is a case in point. Some coals contain 10,000 times the concentration of germanium that occurs in average sedimentary rocks (Vlasov, 1968). It is unlikely, even if chemical processes favoured it, that the flow of ground water could have transported such quantities of germanium into the coal. How then does germanium get there?

Germanium is an element which chemically is distinctly similar to carbon. (It is in the same column and two rows below carbon in Mendeleev's periodic table of the elements.) It forms compounds with other elements similar to those of carbon, and among them, for example, is the compound germane (analogous to methane), GeH_4. Germanium is of course very much rarer in the Solar System and in the Earth than carbon, but in that lower proportion it may often have shared a common history. Methane coming up from deep may have transported germane with it, coming perhaps from the same initial source material, and the dissociation of germane can be expected to be similarly initiated by carbon as the dissociation of methane.

Mercury, uranium, gallium and many other trace elements are also often found concentrated in the coal, much beyond normal sedimentary levels. Again, chemical explanations have been advanced where the chemical reducing action of the coal is invoked, but in most cases of these extreme concentrations it is difficult to see that groundwater or any other agency would have transported enough of the substance through the coal, even if all of it was arrested there.

While most commercially produced coal seams are layered between sedimentary strata, there are many coal deposits in the world that are not. Coal that is interbedded with volcanic lava and without any sediments is known in several volcanic areas, most notably in southwest Greenland (Pedersen and Lam, 1970). It also occurs on the west coast of Greenland in close proximity to large lava-encrusted lumps of metallic iron, not far from mud volcanoes that issue methane and a rock face which frequently has flames burning in its cracks (Henderson, 1969).

Another notable nonsedimentary coal occurrence is in New Brunswick. There a coal called Albertite fills an almost vertical crack that goes through many horizontally bedded sedimentary layers (Hitch-

160 *The origin of petroleum*

cock, 1865). It was mined in the last century, but the difficulty of mining a near-vertical seam caused the operation to be stopped.

Deposits of coal-like materials spanning the whole range from the normal coals that contain some hydrogen and sulphur, oxygen, nitrogen, and so on, to the pure carbon of graphite, exist in many locations in igneous and metamorphic rocks. Most of these deposits have been interpreted as due to the invasion of some biological materials, but some of the deeper levels of graphite have been accepted as due to some internal processes in the Earth. Of course in a sense the much deeper levels of diamond that we know to exist are really no different. We could think of carbon as emanating from deep levels and making its way up, with the tendency to concentrate where it once starts a deposit. This would result in the deposition of diamond, graphite and the various forms of coal vertically stacked above each other, and also in the carbon enrichment and the oxygen depletion in the surface deposits of peat and brown coal.

Carbonaceous shale would fit well into this picture. A dense clay with very small pore spaces is a very good molecular filter. If a layer of such clay overlies a source area of hydrocarbons, it is probable that the heavier hydrocarbon molecules will be removed from the stream and get stuck in the pores of the clay. Eventually this process will result in a clay that has all its pore spaces filled with heavy hydrocarbon molecules, and that is essentially what carbonaceous shale is. The fact that the hydrocarbons in it show a close resemblance and contain many types of molecules similar to those in neighbouring or overlying oil fields would therefore not be surprising at all. Both would be due to the same source materials and the same migration paths, and the similarity between the shale oil and the local petroleum deposit, which is often clearly seen, cannot be used as a proof that the petroleum derived from that shale.

It seems that a lot of clay was laid down in the geologically early Devonian period and therefore many sedimentary deposits have Devonian shale in their lower reaches. If petroleum or natural gas is found somewhere in such a sedimentary region, there is a good probability that an early carbonaceous shale will be somewhere underneath the deposit. While the identification of this shale as the source material for the petroleum or gas would, in our view, be an error, it still is not wrong for the prospector to take its presence seriously: carbonaceous shale at a deep level is a good indicator that the area has been pervaded by hydrocarbons and therefore that oil or gas may be found somewhere in the neighbourhood.

The fossil content of carbonaceous shale is not known to be significantly higher than that of any other type of shale. Whether any of the fossils really deposited unoxidized carbon or were fully oxidized before being buried is not known, and in any case the total mass of unoxidized carbon that would have belonged to the plants or animals whose fossils are found is a very small fraction of the carbonaceous material in the shale.

The distributed forms of coaly or tarry materials found in little specks distributed through many sediments may have a similar origin. Again, this material has been regarded without question as being of biological origin, although no evidence for this viewpoint has come to light. It has been regarded as a possible source material for neighbouring oil or gas, although no explanation has been found for a transportation mechanism from such a diffusely distributed solid source material to a much more concentrated liquid petroleum deposit. Again, it is more likely that the specks of kerogen are caused by the precipitation of carbonaceous material from an upward stream of hydrocarbons in the area, and that any local petroleum was formed by the same stream. The prospector who considers the presence of kerogen as a good indicator for local petroleum is correct, but the interpretation that this is due to the kerogen being the source material for the petroleum is, in our view, incorrect.

We see that in a stream of hydrocarbons there is a general tendency for carbon to form concentrated deposits. The autocatalytic action of carbon deposits causing more deposition of carbon from the stream may be the major factor in the deposition of kerogen and coal. Where petroleum has been deposited to the extent of filling all the local pore spaces in an area, it will obstruct the upward flow of more oil-laden methane. All the upstreaming mix will have to dissolve in the oil, and then some of it will re-evaporate from the top of the layer. It is likely that this process will arrest a large part of the heavier hydrocarbons previously carried in the stream, and that only the lightest components will evaporate again at the top. It is a very fortunate fact for the users of hydrocarbon fuels that both the chemical and the physical properties of carbonaceous deposits have such strong tendencies to make highly concentrated deposits.

The huge deposits of methane hydrates

Large areas of the Earth are permanently below the freezing point of water, and all the pore spaces in the shallow surface layers are then permanently occluded by ice. If methane were commonly seeping up through cracks and fissures from below, then we would expect these areas of permafrost to obstruct this flow, and we would expect to find methane in them generally. In fact this seems to be the case.

A form of water ice combined with methane, termed methane hydrate (which we have already mentioned), has been recognized only in recent years as representing an enormous deposit of hydrocarbons. The permafrost regions of Siberia, Northern Canada and Alaska all have large quantities of methane hydrates buried in the soil. In addition, and on a larger scale still, there are the methane hydrate deposits of the ocean floor, recognized by now in thousands of locations in the Arctic Ocean and in the cold, deep ocean trenches that run along the edges of some of the continents (Kvenvolden and Barnard, 1982) (Fig. 16).

The physical properties of methane hydrates are somewhat different from those of plain water ice. At elevated pressures an addition of methane will raise the freezing point of water ice, so that methane hydrates can be present where plain water would have melted. In deep ocean trenches the temperature may allow the sea water to be liquid but methane-loaded ice to form in the sediments just beneath.

When methane hydrates were not expected, and when this change in the freezing point was not known, they were readily overlooked. If a sample of ocean floor material was brought up, it was usually not maintained at the pressure it had on the ocean floor. The methane hydrates therefore melted and the methane escaped. Similarly in permafrost samples, the procedures that were followed were inadequate to preserve the methane. Only in recent times, when all this was understood, was the sampling procedure done with sufficient care, and only then did it become clear how vast was the Earth's store of this substance. Much of the deep ocean water is just above freezing, because of circulation of the cold water that sinks down in the polar regions. The combination of temperature and pressure will maintain that form of methane-loaded water ice in the sea-floor sediments. Thousands of core samples have been taken from the ocean trenches in the Pacific, along Peru and Guatemala, and from many points from the shallower but colder Arctic Ocean,

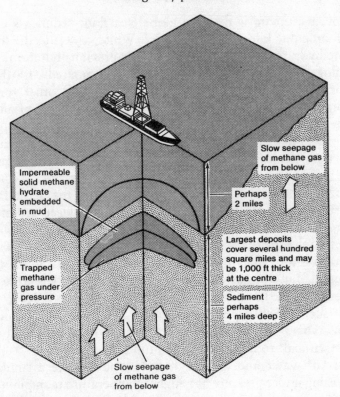

Impermeable
solid methane
hydrate
embedded
in mud

Slow seepage
of methane gas
from below

Perhaps
2 miles

Trapped
methane
gas under
pressure

Largest deposits
cover several hundred
square miles and may
be 1,000 ft thick
at the centre

Sediment
perhaps
4 miles deep

Slow seepage
of methane gas
from below

Fig. 16: Methane hydrates on the ocean floor.
Methane hydrates (water-methane ices) have been discovered in very many areas
of the ocean floor, chiefly by ship-borne drilling from Glomar Challenger. These
deposits suggest that methane seepage from below is an extremely widespread
phenomenon, since these ices exist almost everywhere where they would be stable.

and it seems that wherever the pressure and temperature conditions
would have allowed methane hydrates to form, there is a high
probability that they can be discovered.

Where methane hydrates have been identified, it was frequently
possible to recognize their level in the ocean floor deposits by a
particular reflection of sonic signals from the type of sonar apparatus
that is used for hunting submarines. This reflecting horizon does not
follow the layering of the ocean deposits, but instead follows the
surface at which pressure and temperature would just allow hydrates
to form. In many places this surface cuts at an angle through the
deposits, and it therefore seems pretty certain that it is correctly
identified. On the basis of extrapolating from the deposits discovered
by actual cores and the extension by the sonic technique, absolutely
gigantic estimates of the total methane content have been produced.

One Soviet estimate is that the global sea floor sediments contain one billion cubic kilometres of gas (cf. White, 1985) in the form of methane hydrates, which, in terms of our previous notation, equals 0.14 kilogram per square centimetre, to be compared with the estimate of 3 kilograms per square centimetre of all other forms of unoxidized sedimentary carbon. It far exceeds the total of all other estimated sources of natural gas.

If all this methane had been produced from biological materials, much larger deposits would have to be invoked, even larger than for the production of all the other hydrocarbons and of coal. Yet some of these hydrates on the ocean floor, or in the permafrost of the northern regions, do not overlie enormously deep sedimentary deposits. Their presence in such large quantity fits readily with the assumption that methane seeping up from deep levels is a widespread phenomenon and that the quantities involved are large enough to have supplied all the carbon on the Earth's surface. The widespread occurrence of methane hydrates would have been predicted on the basis of such a theory, and had they not been found, it would have been an embarrassment. As it is, they seem to be there, and it would be quite difficult to account for them in any other way.

A mix of water and carbon dioxide would make a similar ice, also freezing under pressure at a higher temperature than plain water ice. It is remarkable that carbon dioxide hydrates are much less common than methane hydrates. If sediments were pulled under to a sufficient depth to cook out methane from the organic content, it would certainly also cook out carbon dioxide, and if it was associated with sedimentary carbonates, the quantity of carbon dioxide cooked out should usually exceed by far the quantities of methane. If big slabs of sediments were pulled under to great depth, as has been suggested, and if the gas production from them could make its way towards the surface, then one would expect on an average very much more carbon dioxide than methane or other hydrocarbons to come up. The regions of hydrate occurrence are a good testing ground, and the evidence certainly suggests that methane is the dominant gas that comes up from below in all these areas.

While there are still conflicting opinions on how widespread the methane hydrates really are, and whether they are really present in all locations where pressure and temperature would allow them to exist, it seems certain that carbon dioxide hydrates are more rare. It is interesting to compare this situation with that in the hot pathways of outgassing. There carbon dioxide is vastly more common than methane, as we have seen in volcanoes, and also in the great ocean rifts. The hydrates thus seem to confirm the suggestion that on cool pathways methane dominates, while carbon dioxide dominates in the hot regions. Mud volcanoes gave us the same suggestion, and we also saw that helium of deep origin in volcanic vents is associated with carbon dioxide, but in nonvolcanic vents it is associated with methane.

There has of course been some discussion whether methane in this hydrate form can be used commercially. It seems that it is possible, but by no means easy. A large amount of heat has to be supplied to the deposits to liberate the methane, and it is of course expensive and difficult to produce so much heat and to distribute it over large volumes.

However, the methane hydrates provide an excellent barrier to the escape of methane, by freezing up all the pore spaces. As a result, we find accumulations of free methane below, and these may be of greater commercial significance than the hydrates themselves.

Perhaps if we ever run out of more convenient sources of methane, we may have to try to melt it out of hydrates. But for the present, I believe that large quantities remain to be discovered in reservoirs from which production is more economical.

10

CONCLUSIONS

Oil and gas are found in very diverse settings in rocks spanning all ages, in mountainous and in flat regions, in the middle of continental plates and on their edges, in sedimentary and in volcanic rock, and even in cracks of the ocean floor. It seems that the only generalization that can be made is that there is no type of geological setting which is not somewhere associated with gaseous, liquid or solid hydrocarbons.

This is not to deny that there are strongly preferred settings. The edges of the continental blocks, the fold-belts along which mountain ranges have been thrust up, the major rifts in the continental or oceanic crust, and the vicinity of arcs of volcanoes, such as the oceanic island arcs or arcuate volcanic lines on continents, all these are certainly preferred regions. Such large-scale tectonic features may be favoured either because they represent lines of weakness in the crust, along which the outgassing process is facilitated, or they may represent lines along which the underlying mantle of the Earth is particularly rich in volatile substances or particularly active in driving them out.

Most petroliferous areas show signs of hydrocarbon seepage on the surface, and it is clear that the search for such seepage is a valuable method of prospecting. Most oil in the world has been found by searching in the vicinity of obvious surface oil seeps. Prospecting for gas can similarly be done through the identification of gas seeps, but there some instrumentation is usually required. Most gas seepage seems to go unnoticed, except if it is under water and bubbles are noted, or if some particular reason is advanced for an instrumented investigation of an area.

In addition to a direct detection of the gas, one can also observe chemical effects in the surface soil that are recognized to have resulted from methane seepage, or to be generally associated with it. For the explanation of these effects we have to suppose that enormous quantities of gas have escaped and are still escaping, and in many areas we have to consider that far larger quantities have escaped than can now be found in any underlying reservoirs. If we suppose, as we do, that oil was transported and implaced by solution in methane, then certainly much more methane than oil must have come up. Since it is much more mobile,.we would certainly expect methane surface seepage to exceed greatly the seepage of oil. Modern instrumented investigations on land, sonar detections of bubbles and chemical detection at sea, and the discovery of the widespread presence of methane hydrates on the ocean floor and in permafrost confirm this conclusion.

All this raises two major questions. Firstly, how will this outlook affect the techniques and strategies for prospecting for fuels in future and, secondly, how does it affect the long range energy outlook?

Prospecting techniques and strategies

If hydrocarbon fluids come up from great depth, forcing their way up because of their buoyancy relative to the rock, then we have to think of the hydrocarbon reservoirs that we seek in a completely new way. These would not now be locations where locally available fluids happened to have become trapped under a dense layer. Instead, they would be places where a stream, coming from a high pressure source below, has been dammed up sufficiently to fill and even to expand pore spaces and thus create and maintain a reservoir. The flow rate of such a stream would be completely defined by the situation at a depth much below the levels we explore; whatever fluids are commencing their ascent from down there will eventually appear on top. Variations in the properties of the rock at shallow levels cannot permanently alter the rate of delivery, although such variations may alter the amounts stored on the way. We may compare this with the flow of an (idealised) river which derives all its waters from a supply in the mountains, and then flows, without further supply, over the lowlands and out to sea. The long-time average of the amount reaching the sea has to equal the rate of supply; nothing that happens in the plains can permanently alter that.

What a dam on the river can do, however, is to create a storage reservoir upstream from it. Eventually, that dam must overflow and the flow-rate to the sea will then return to the value given by the rate of supply in the mountains. In a similar way, a less permeable layer in the outer crust will dam up an ascending stream of hydro-carbons and create a reservoir below. But, in the long run, the reservoir must overfill and the flow-rate to the surface must return to the value that equals the rate of supply. On this basis it is clear that oil and gas fields will, in general, have seepage of oil and gas above them at the surface. This indeed appears to be the case. Regions that have been well explored by the drill show the highest seepage rates closely related to the areas that have the highest content in the reservoirs below. Gas fields frequently show surface gas seepage in characteristic halo patterns around their perimeters, showing the spillage of an oversupply.

If the oil and gas had been generated locally in the sediments, would we then not have to expect that many regions had a reservoir capacity exceeding the supply? Should we then not have found much oil and gas with no surface seepage above? If the quality of a caprock was a feature limiting the availability, should we then not have seen oil and gas fields characterized by a particularly low rate of seepage above them? And, in that case, why would a gas of totally unrelated origin, like helium or mercury vapour, show surface seepage patterns closely coinciding with those of the hydrocarbons? If, however, the picture of a stream applies, then the observed relationships are to be expected. Reservoirs are constantly being overfilled; unrelated volatiles join the stream and are brought up with it. The strategy of prospecting for oil and gas then takes the form of first looking for areas where the surface seepage is pronounced, where the flow appears to be strong. In such areas it will have taken less of a hindrance to the stream for a reservoir to have been created. Where the stream is less strong, a more perfect barrier would be required.

At shallow levels, a tight caprock will still be an extra requirement to dam up the stream and hold down a hydrocarbon reservoir, especially one of gas, but the requirement of an organic deposit, a "source rock", will have fallen away. At deeper levels, below what we have termed the "critical layer", even the requirement of a caprock will have fallen away. Without these two requirements of a biological deposit and a competent caprock, the estimates of the quantities that may be found will increase greatly. The levels below the critical layer may occasionally contain oil, but gas is far more common there. The expectation is therefore that gas will become the major fuel for a long time.

Surface seepage of gas is common in many parts of the world, and prospecting for it is not difficult with modern instrumentation. After that, it is necessary to locate suitable reservoir rocks at levels below the critical layer. These must be either rocks that have an intrinsic porosity which may be held open by the gas pressure there, or rocks which will be fractured by the high pressure gas to form a connected and well-distributed network of pores. Sediments below 15,000 feet will still be a prime target, not because they produce gas, but because they frequently contain a finely divided porosity which is good as a reservoir. Furthermore, sedimentary rock tends to be softer than igneous, and the critical layer is therefore likely to be shallower. Not only large sedimentary basins that reach to great depth will come under consideration, but also many rifts in which sediments have accumulated. In many rift valleys the depth of the sediments is not yet known, and even quite narrow rift valleys could contain large quantities of gas at deep levels.

The ordinary crystalline basement rocks are generally too dense to be reservoirs. There may be cracks, and they may often contain methane, as they do, for example, in the super-deep hole in the Kola Peninsula at a depth of 12 kilometres (Kozlovsky, 1984), but for a commercial flow-rate into a well-bore an occasional crack is not adequate. Such rocks will come under consideration where there is some reason to believe that they have been smashed. Severe faulting may be one such reason, and the search for gas in such zones may be profitable. Ancient meteorite impact regions, where very large meteorites hit long ago and created vast volumes of smashed basement rock, are the most striking of the new possibilities.

At the deeper levels, in the 15,000-30,000 feet range (5-10 kilometres), the gas is at a high pressure, and therefore enormously more dense than at shallow levels. At 21,000 feet depth one cubic foot of pore-space may contain 2000 cubic feet of gas, measured at atmospheric pressure. This fact, and the huge pressure differentials when the gas is tapped, often result in a very large total production from one well, and also in high flow-rates, even in rocks whose permeability would be considered inadequate at shallow levels.

Many deep wells have thus been enormously successful and have repaid the drilling costs in a short time, especially in Oklahoma where the art of deep drilling for gas is most advanced. The expectation is that once we know that there is a lot of gas to be obtained, and when industry has changed over to gas from other less desirable fuels, deep gas, with its many advantages, will be perfectly competitive.

Many areas of the world will come under consideration where, on the conventional viewpoint, there were no expectations. The porous, volcanic rocks of Japan are being explored now and already substantial quantities have been found in circumstances which would seem most unlikely on the basis of a biological origin. Perhaps very much larger quantities will be found there when the present success encourages deeper drilling. The Japanese island of Hokkaido seems particularly promising, as there is much evidence that the line of violent outgassing of the Kurile Islands to the north continues into Hokkaido, and as the large deposits of coal there are so permeated by gas that it can no longer be mined. It has become clear that much more methane is there than could have been maintained in the coal over its age.

The techniques of chemical surface prospecting have already been demonstrated to work very well for the discovery of the past and present seepage that indicates an underlying reservoir. The problem with these techniques was merely that one could not understand why they would work; the quantities of gas that must have escaped to give the observed indications seemed improbably large and therefore there was much hesitation in accepting techniques that seemed unreasonable. On the basis of the present picture, these large quantities of escaping gas can be expected to be quite real and a search by these techniques is then quite realistic (Donovan *et al.*, 1974; Donovan, 1974; Duchscherer, 1981, 1984).

In addition to leaving chemical traces in the ground, upwelling methane also tends to cause iron in the ground to be chemically reduced, almost certainly by the action of bacteria, and the magnetic patterns that result from this can be identified. Gases tend to stream through the ground in a very uneven way, holding open pathways that they have created; it follows from this that in an otherwise uniform terrain, a very spotty magnetic signature is a good first indication of methane seepage (Donovan *et al.*, 1979). Airborne magnetometry is a very quick and comparatively inexpensive method of obtaining a first indication of methane seepage of an area, and it can then be followed up by verification with ground-based instruments. On the ground both the direct detection of soil methane and the detection of particular carbonates that have been created from it, sometimes in gigantic quantities, would be the next step. The distribution of certain trace metals which appear to be carried up by hydrocarbons gives a further indication (Duchscherer, 1984). These methods will certify an area as being one in which methane is coming up, and the only remaining question is then whether a porosity domain at deeper levels is likely to exist there.

The Swedish experiment

Despite the apparent strength of all these arguments, it was clear that the firmly established viewpoint of the biological origin of all of the Earth's hydrocarbons could not easily be displaced. An example of a large hydrocarbon occurrence in an area where there was little or no chance that biological deposits could have contributed would clearly have the greatest impact. The ancient granitic shield of Scandinavia seemed a prime candidate. If hydrocarbons seep up from the mantle below, then the granite of Sweden appears to be resting on an area of the mantle that is particularly rich in hydrocarbons. The Norwegian trench on one side with its numerous oil and gas fields and the many hydrocarbon occurrences, as well as some commercial production to the east and south, could be understood in that way. It was also known that many cracks in the granite of Sweden have deposits of tar in them which may also be due to hydrocarbon seepage from below. If any large region of porosity existed in the granitic rock of Sweden, it might be expected to contain gas and oil.

It was for this reason that I made contact with the Swedish State Power Board, where an influential group (under the direction of Dr Tord Lindbo) was willing to pursue these ideas. Among the areas where porous rock might be found, the outstanding candidate was Europe's largest meteorite impact crater, situated in Central Sweden and known as the Siljan Ring. When investigations were started there it was quickly discovered that much evidence already existed, pointing to just this structure as an area of seepage of both gas and oil. Farmers' water wells had long been known in the area to produce combustible gases, and numerous oil seeps were noted in the small sedimentary deposits of the ring-shaped depression that marks the crater.

After three years of investigation it became quite clear that the area was indeed outstanding for all the effects that are known in other oil and gas areas of the world to be associated with the seepage of gas. Anomalously high levels of methane were found in the numerous lakes and ponds of the area. Carbonates of a kind recognised to be derived from the local oxidation of upwelling methane were found in characteristic patterns in the surface soil, together with trace metals that also usually accompany such seepage. Methane-derived carbonates were also found in the seven probing holes that have been drilled to a depth of approximately 500 metres.

Hydrogen, methane, ethane, ethylene and propane were found in all these holes, usually in small concentrations only, but occasionally in concentrations approaching combustible levels. A gravity survey over the area showed a marked low-gravity anomaly, similar to such anomalies known over other large impact craters, and interpreted as due to substantial porosity in the smashed rock. In the area of the maximum negative gravity anomaly, seismic profiles showed several strongly reflecting horizons, stacked one above the other, implying changes in density or sound speed at these levels. Porosity of course has a very major effect on the sound speed, and one may therefore interpret these reflections as due to mineralization processes that have cemented up the pore spaces at particular levels, leaving different fluid pressures and different porosity proportions above and below each such layer. It is this information as well as some other corroborating items that suggest that fluids are trapped in this region, and that hydrocarbons are among them.

The total volume of pore space, as indicated by the gravity measurements, is very large. If filled with gas the area would be a gas field of world class. Of course the pore spaces may be just filled with water, but then one would not understand why the near surface showed such strong indications of gas and oil. Even a small fraction of the pore spaces filled with hydrocarbons could make it a commercially very significant field.

The Swedish State Power Board has now organized a drilling program to drill to a depth of between 15,000-23,000 feet. The American Gas Research Institute is making a significant financial contribution, on the basis that the Swedish experiment will tell us much about the question of the origin of hydrocarbons and, with that, about the global availability of gas in hitherto unsuspected regions.

At the time of writing, very significant shows of gas have been detected on the way down at depths below three kilometres in the granite, already making clear that gas is indeed streaming up through the broken rock. The quantities of methane-derived carbonates have also been observed to be so large that any local or surface origin can be firmly ruled out. That large quantities of gas are indeed somewhere below seems clear, and the only uncertainty at this time is whether a sufficiently good zone of containment and porosity will be encountered at the deeper levels.

The world energy future: the methane age

If deep gas is plentiful and can be discovered in many parts of the world and in many geological settings, the economic outlook would be greatly improved and many countries would breathe more freely when they are relieved from the obligation to build or expand a nuclear power industry in a hurry, or from the need to depend on imported fuels. The economic and strategic world picture would change when the ownership or the defence of the principal fuel-supplying regions becomes less critical. Oil will of course continue to be important as a transportation fuel and will be used for as long as possible for that purpose. The stationary use of fuels will shift over in many or most parts of the world to natural gas, once it is produced and the pipeline infrastructure for its distribution is built. From the environmental point of view, this will be a substantial advantage, since gas is by far the most benign of all fuels. Sulphur, if it is present to start with, can be taken out very completely before burning and therefore whatever acid rain problem is due to sulphur would be greatly diminished. The numerous environmental problems associated with the mining, transportation and burning of coal would not have to be faced. If the building up of atmospheric carbon dioxide, which has been observed, is harmful to the global climate, then the burning of methane is also far preferable since, for the same energy produced, it delivers only half as much carbon dioxide as coal or about a third less than oil.

What happens when the oil finally does run out? Can the world make itself totally dependent on gas?

There is no real barrier against the conversion of methane to liquid fuels. Some commercial processes already exist and other better ones will almost certainly be invented. At the present time, natural gas, in compressed cylinders, is used on quite a large scale in automobiles, where it is found to be much safer than liquid fuels and really very satisfactory from many other points of view. Perhaps this usage will also be greatly expanded.

How long will the natural gas last? Will the Earth replenish some of the reservoirs as we drain them? Is there a virtually limitless supply at the deeper levels which we are just now only beginning to tap?

It is hard to say whether, on the basis outlined here, the amount of gas we can tap with present drilling techniques will last 100 years or 500 years. Certainly we shall be exhausting most accessible reservoirs faster than nature would refill them, although here or there a significant refilling process may be at work, and indeed seems to have been noted already in some cases. We may hit on some deep reservoirs that are virtually inexhaustable, where the natural refilling from below occurs as fast as we can withdraw the gas. But the long term outlook has to be that it will be very many years before all the reservoirs down to 30,000 feet have been exhausted, and by then one may expect that totally different and commercially viable sources of energy will have been discovered, or drilling techniques will have been improved so that the next, deeper horizon of even higher pressure gas can be tackled.

APPENDIX I
The carbon isotopes

Carbon has two stable isotopes: carbon-12 (6 protons, 6 neutrons), and carbon-13 (6 protons, 7 neutrons). The natural carbon on the Earth contains predominantly carbon-12, and carbon-13 is mixed in at a level of about 1 percent. This mixing ratio must have been determined in the nuclear processes in stars that cooked up the elements and eventually supplied them to form the planets. There are no processes that could occur on the planets that would be able to change this ratio greatly. It is only small variations that can be produced, not by any effects on the nuclei themselves, but only by processes that show a slight preference and select in favour of either the light or the heavy isotope.

The interest in the carbon isotopes for geological purposes arises just because of these slight variations, which can be found in carbon from different natural sources. A precise measurement of the mixing ratio can thus reveal something about the history that any particular carbonaceous material may have suffered. Thus the study of the distribution of the carbon isotopes in relation to petroleum and natural gas has a very extensive literature; we shall discuss here only one aspect of it: can isotope measurements determine whether a hydrocarbon compound is derived from biological material or is primordial? Since many petroleum geologists have considered that such a distinction can be made, and that petroleum and natural gas appear, on that basis, to be usually of biological origin, it is clear that we must discuss this aspect here.

A selection, enriching the product in one or other isotope, is usually referred to as a process of "fractionation". The resulting "fractionated" material is called "isotopically light" or "isotopically heavey" as the case may be.

Measurements of the slight variations in the carbon isotope ratio in different samples is usually not done in absolute terms, but by comparison with a norm. The small departures from this norm are then the quantities noted. The norm that has been selected for this purpose is a marine carbonate rock called Pee Dee Belemnite, or PDB, which, for its carbon isotope value, lies about in the middle of the distribution of all the marine carbonates. The measurements are then quoted as the departure of the carbon-13 content of the sample from that of the norm, and the figure is usually given in parts per thousand (per mil), and referred to as the $\delta^{13}C$ value of the sample. Thus if the norm were precisely 1 percent of carbon-13 (it isn't), then a content of 1.001 percent would be denoted as a $\delta^{13}C$ value of +1 per mil, or a content of 0.999 as −1 per mil.

The unoxidized carbon in plants, traditionally referred to as "organic carbon" – in contrast to oxidized or "inorganic carbon" – derives from atmospheric carbon dioxide by the process of photosynthesis. The energy of sunlight dissociates the carbon from the oxygen in this complex process, and there the light isotope is slightly favoured. As a result this carbon is slightly depleted in carbon-13 relative to atmospheric carbon dioxide and most other forms of oxidized carbon. The effect, though small in absolute terms, is larger than that occurring in any other single non-biological chemical process in nature. When it was found that most of the deposits of unoxidized carbon, like petroleum, methane, coal and kerogen, showed also a marked depletion of carbon-13, it was considered that this confirmed their biological origin.

Plant "organic" carbon is generally in the range of −10 to −35 per mil (PDB standard); the atmospheric carbon dioxide is at −6. Marine carbonates, laid down from atmospheric carbon dioxide dissolved in ocean water, range from about +5 to −5 (the PDB norm having been chosen in the centre of the range), and thus evidently a fractionation of about 6 per mil occurs in favour of the heavy isotope during the carbonate formation.

In the production of methane from plant debris – a process that undoubtedly occurs – a further fractionation takes place that again favours the light isotope, and plant-derived methane is therefore isotopically even lighter and plots at −50 to −80 per mil. In the literature we now find that some arbitrary division has generally been made around a carbon isotope ratio of −30 per mil, below which any carbon is regarded as of biological origin and above which it is taken to be from some other source.

However, there is no clear division in the actual data. Carbonaceous materials span the range from −110 to +20 on the Earth and they span an even larger range in the carbonaceous meteorites. There is no natural dividing point in the data, and the choice of a particular figure somewhere in this continuous distribution, for making the distinction between organic and inorganic origin seems

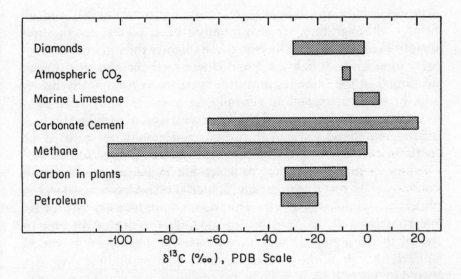

Fig. 17: Distribution of the ratio of the stable isotopes carbon-13 and carbon-12 in different terrestrial materials (PDB scale).
Methane and carbonate cements span a much larger range of isotope ratios than all other forms of terrestrial carbon.

very arbitrary. The question is of course very simply whether other fractionation processes than biology can select in favour of the light isotope by similarly large amounts.

A look at the distribution of the carbon isotope ratio in different forms of carbon gives immediately a strong suggestion. (Fig 17). The atmospheric carbon dioxide from which marine carbonates (limestone and dolomite) have been deposited throughout geological time seems to have had a remarkably constant isotopic value, so that nearly all these carbonates fall into the range of −5 to +5. Carbon in plants is isotopically light, as we have said, ranging from −10 to −35. Petroleum has a fairly narrow range from −20 to −38. It is just methane and the carbonate (calcite) cements in the rocks which span a very much wider range. That itself of course suggests that the carbonate cements are generally produced from methane, as we have already discussed. The shift of the carbonate cement by between 20 and 40 per mil points to the heavier side then looks like a fractionation that occurs when methane is oxidized in the ground and then combined with calcium oxide to produce the carbonate cement.

Everything we know about these carbonate cements points to such a process. They are found in great quantity overlying gas and oil fields – although they are also found in other places, but in lesser quantities. They are usually isotopically lighter than marine carbonates, sometimes as light as −65. Where methane and this cement are found in the same location, the methane is isotopically lighter still, by between 20 and 60 per mil.

We have already discussed Kudryavtsev's rule, according to which any region that shows hydrocarbons tends to have some at all levels below, down into the basement. We may now add to that "Galimov's Rule", according to which methane in any such vertical column tends to be isotopically lighter the shallower the level at which it is sampled (Fig. 18). This, again, appears to be true in the great majority of the cases investigated, irrespective of the type or age of the formation from which the sample was taken. It is most unlikely that in all those cases methane from two different sources mixed in this fashion; a much better explanation is that a *progressive fractionation* of the methane takes place as it migrates upwards. Some of it is always being lost to oxidation, ending up as carbon dioxide and from that as calcite cement in many cases. This always prefers the heavy isotope (as isotope exchange calculations would indicate that it should), and so the remaining methane gets isotopically lighter and lighter. At each level the calcite thus derives from the fractionated methane, and it also will become lighter, tracking the methane, but always remaining heavier than the methane at that level.

Progressive fractionation is a very important process, because it can drive the remaining material to a very much greater fractionation than could be done by any single chemical step. It is the technique used for commercial isotope separation, where extremely large fractionation factors are required. In our case, there are two effects that work in the same sense, helping to create the result. One is the tendency for the oxide to bind slightly more tightly with the heavier carbon isotope, and in equilibrium conditions at a low temperature the heavy isotope will therefore be enriched in the oxide. The other effect is the diffusion speed, which for methane with the heavier isotope is 30 per mil slower than for the light one. This means that in any circumstance where the methane is diffusing through a barrier, the heavy isotope will reach a concentration which is enhanced by 30 per mil. Oxidation occurring in those circumstances may then be favouring the heavy isotope by an extra 30 per mil. (A more detailed

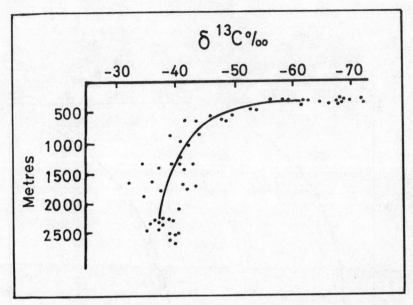

Fig. 18: Carbon isotope ratios of methane plotted against depth of occurrence (from Galimov, 1969). Although there is much variability in this relationship, it is almost always true that where methane is found at different levels in the same area, it is isotopically lighter (less carbon-13) the shallower the level.

discussion of this point is required to see under precisely what conditions this additional fractionation effect will come into play.)

The possibility of such a progressive fractionation occurring in natural circumstances seems not to have been considered, and without it one was naturally inclined to look for the largest effects that were known in single steps, to account for the large carbon-13 depletions that were often seen in the hydrocarbons, especially in methane, and those certainly occurred chiefly in biologically mediated chemical reactions. With the strong evidence for progressive fractionation we now have, this is clearly a contender for the explanation (Fig. 19).

It has been argued that the anomalous, isotopically light carbonates can be found in locations where there are no known hydrocarbon reservoirs, and indeed in igneous and metamorphic rock where conventional wisdom denied the possibility of hydrocarbon deposits. Therefore, it was said, this carbonate must have some other, as yet unknown derivation. In the context of our discussion, this is just what would be expected: the outgassing of methane from deep levels is a common process, occurring to a varying extent, but in many or

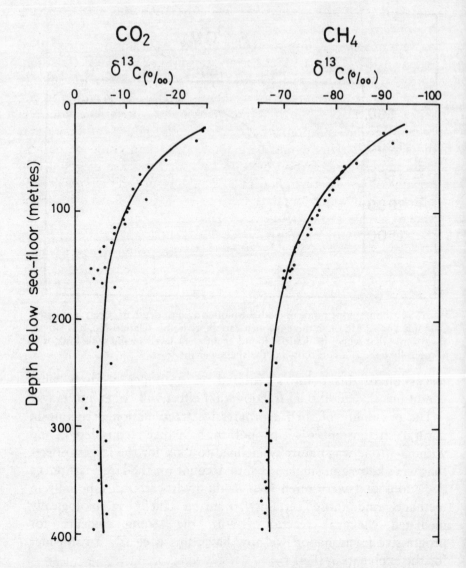

Fig. 19: Comparison of the carbon isotope ratios of methane and co-existing carbon dioxide in ocean floor sediments (from Galimov and Kvenvolden, 1983). The carbon isotope ratios of the two gases seem to follow the same depth dependence, but with the CO_2 always isotopically heavier than the methane. This is what would be expected if progressive fractionation were happening, with the CO_2 produced (probably by bacterial oxidation) from upwelling methane. Both the methane and the CO_2 produced from it would then become isotopically lighter on the way up, but the CO_2 would be heavier by a constant amount than the methane from which it was derived. (This is not the interpretation given by the authors of the article.)

most areas of the globe. The widespread presence of methane hydrates tells a similar story.

With this explanation we would expect to see another effect also: where methane appears that has not yet suffered much fractionation, as for example in a fast stream from deep, it may be isotopically heavy, in the same range as diamonds, between -5 and -10 per mil on the PDB scale. If it produces now a carbonate which is perhaps 30 per mil points heavier, we could find a carbonate that plots at $+20$. This would be a lot heavier than could be derived from atmospheric carbon dioxide, and in that sense it would also be anomalous. There are indeed examples of just this. In the granite of Sweden, where many data have been obtained both of methane and of calcite cements, methane has been seen as heavy as -20 (heavier than any biologically generated one) and calcite as heavy as $+18$ has been seen. In the mud-volcano region of Azerbaijahn (Southern Soviet Union) carbon dioxide as heavy as $+23$ has been seen. (Valyayev and Grinchenko, 1985; this paper discusses many aspects of interest to the present subject.)

Many complex isotope fractionation effects may be involved before we see the final product, and most of them are still not fully understood. The possibility of a progressive, cumulative effect means that the end-product can show a far larger fractionation than that in any one step, and for this reason one cannot judge a hydrocarbon to be biogenic just on the basis of being isotopically light.

There is one aspect in the discussion of the ultimate origin of the Earth's surface carbon where the carbon isotope evidence appears to be quite decisive, and we shall briefly direct our attention to that.

Is surface carbon mainly primordial or is it recycled?

We have discussed the large quantity of carbon on the surface and in the near surface rocks which is presumed to have come originally in a fluid form from deeper levels. The question arises whether this carbon outgassing process occurred early on in the history of the Earth and then ceased, or whether carbon outgassing has been a continuous process to this day.

The quantity of carbon in the atmosphere and ocean (those two reservoirs interchange carbon dioxide fairly quickly) is 8 grams per square centimetre averaged over the Earth, while the amount in the accessible sediments is estimated to be of the order of 20 kilograms per square centimetre or 2500 times as much. It is clear that the atmospheric-oceanic carbon dioxide would supply the sedimentation rate of carbonates and unoxidized carbon for a very short time only. Since this sedimentation derives principally from atmospheric-oceanic carbon dioxide, a continuous resupply of this must be envisioned. The question therefore arises whether such resupply is due to a continuing outgassing process from primordial materials in the Earth, or whether it is from early sediments that are being subducted and which, upon heating, send up carbon-bearing fluids.

If recycling had been responsible, then one would conclude that very large quantities of carbon had to be laid down in early times, namely at least all the carbon which we now find in all the sediments, and one would therefore expect to find ancient sediments particularly rich in carbon, compared with more recent ones. This does not appear to be the case, and in fact the opposite has been stated, namely that early sediments contain a smaller proportion of carbon relative to other minerals than the more recent ones.

But perhaps quantitative arguments are not very secure, and the sampling of early sediments is not representative. The carbon isotope data, however, suggest that recycling is unlikely to have been the major source of the continuous carbon supply.

Marine carbonates show no significant change in the carbon isotope ratio from early Archean times to the present. Yet in recycling there would always be a loss of isotopically light carbon in the various forms of unoxidized carbon deposits that are in the sediments. These range in deposits of all ages from approximately −30 per mil to −60 per mil. Most of that carbon will end up as a very stable and permanent solid, namely graphite, in the process of subduction and heating, and it is difficult to see that much of it could ever be returned in a fluid form back to the surface. The carbonates that are subducted, on the other hand, may be taken to a regime of temperature and pressure where they dissociate back into carbon dioxide and metal oxides. It would therefore appear that each recycling process would abstract and not return some isotopically lighter carbon, and therefore the atmospheric carbon dioxide supply would be driven towards the heavier side in each such cycle. If the ratio of unoxidized to oxidized carbon in sediments is 1:4, we might judge that one recycling step would result in an atmospheric carbon dioxide which is between 5 and 10 per mil heavier than the mean of the subducted carbon. If recycling had occurred several times, this isotopic fractionation would be cumulative and therefore constitute a very large effect.

It seems therefore most improbable that recycling could have continuously supplied carbon without progressively fractionating it. The constancy of the carbonates throughout the geological record therefore indicates that they do not represent recycled material, but rather that a continuous and unchanging source material is responsible.

This argument does not make a case against the subduction of sediments, only against the reappearance on the surface of the carbon. It is perfectly possible that both the carbonates and the unoxidized carbon in sediments are taken into a regime of temperature and pressure where neither of them is turned into volatile carbon. The form of calcium carbonate known as carbonatites is known to derive from deep levels, and therefore demonstrates that stable carbonates can exist there.

The remarkable constancy of the carbon isotope ratio of the marine carbonates over all the time of the geological record poses a further problem. If the deposits of unoxidized carbon in the sediments were really derived mainly from plant material that was buried without being oxidized, then any fluctuation in the rate of such deposition should show itself in a change of the isotopic value of the atmospheric carbon dioxide, and therefore of the carbonates laid down from it. Unoxidized carbon in the sediments is thought to account for about a quarter of all deposited carbon, the rest being in the carbonates. Buried plant material would always take down fractionated carbon, isotopically much lighter than the mean, and this process would therefore drive the atmospheric and with it the carbonate carbon to the heavier side. In steady conditions we would not see an effect, since we do not know the precise value of the carbon that replenishes the atmosphere; but if at any time there was a large change in the rate of burial of plant material, this should show up in a change of the isotopic value of the carbonates laid down at that time.

One would have thought that the advent of land plants in Silurian times would have created a great increase in the rate of burial of plant material. It is true that there may have been a lot of marine plants before, but still, the difference between desert continents and continents covered with dense vegetation should surely have made a big difference. We would therefore have expected to see a marked step towards heavier carbonates accompany this proliferation of plants, a step that should be at least 5 per mil, if the quantity of buried plant carbon had been doubled at that time. No such step is found. An explanation that the replenishing carbon supply had undergone a compensating change is difficult to accept.

The problem disappears if we take the view that unoxidized *plant* carbon in the sediments is only a small fraction of all the unoxidized carbon deposits; that most of that carbon derives from the upwelling primordial materials, even if it is found with much biological contamination. In that case there would have been no great change in the deposition of carbon due to the advent of land plants, all plant material making only a small contribution. The supply of primordial carbon into the sediments, both in oxidized and in unoxidized form, could then have been quite continuous, unaffected by any developments at the surface.

Appendix 2
Speculations about the deep interior of the Earth

The deep levels below the crust are not at all well understood yet. We know that the outer mantle is of uneven composition. We have good reason to suspect that materials like the carbonaceous chondrites have made a contribution and we now believe that massive outgassing from such levels has taken place over geological time. We do not know the driving forces that constantly distort and push the overlying crust or why this activity has not died out in the great age of the Earth. The modern view of plate tectonics merely ascribes phenomena to the movement of plates, to the expansion of some surface areas and the contraction of others, and models of the underlying patterns of motion are offered, but the cause of these motions is still a puzzle.

Some authors have favoured the idea that the entire mantle of the Earth should be thought of as a very viscous medium, rather than a rigid one, and that the internal generation of heat by radioactivity would cause this entire mantle to break into a convection pattern, turning it all over perhaps a few times in the age of the Earth. The crust will then be dragged in certain ways by this motion of the mantle.

I do not think that a wholesale convection of the mantle can be expected. Even a very slightly uneven composition, such as a slight variation in the proportion of iron to rock, would cause larger density differences than the heat could produce. If the material was soft enough to convect, it would merely tend to move so as to bring the denser materials down and the lighter ones up and, after it had done that, no temperature differences could be expected to reverse that configuration.

This is not to say that motions in the mantle are impossible or that they could not drive a motion of the crust. It is only that the general overturning of the entire mantle seems unlikely. The material of the mantle is almost certainly in the condition where a comparatively slight increase of temperature would make it much more fluid, or a slight decrease would make it much stiffer. Such a material would always tend to concentrate any motion to those locations that produced or brought in more heat. The regions where this was not the case would "freeze up". Very uneven patterns of motion must therefore be expected. Fluid dynamics in a medium of these properties has not yet been investigated and at the moment we can only speculate what forms such motions might take. It would be my guess that much of the mantle would be essentially stiff and held so by its intrinsic density differences, and that some motions would occur along surfaces threading through the mantle, along which the temperature was maintained at a slightly higher level, and where the material was therefore much more fluid.

What could be the sources of an uneven heat distribution and why would this situation have maintained itself for long?

In an Earth accreted from solids we can expect an uneven distribution not only of the long-lived radioactive elements, but also of the chemical energy sources. Chemical energy has been largely ignored as a significant heat source in the Earth, but it might well be important, especially in relation to the generation of movement. If we now believe that carbon is fairly abundant at deeper levels, and we certainly think that iron is, then the question immediately arises whether the chemical reduction of iron could be taking place basically in the same manner as the production of iron from iron ore and coal. In material like that of the carbonaceous meteorites, the iron is in a highly oxidized form and the carbon is largely unoxidized. If that material were heated, the reaction that would take place would take the oxygen from the iron to the carbon (a consideration first pointed out to me by J. M. Knudsen). The consequence of this reaction would be the evolution of some heat, the production of carbon monoxide and carbon dioxide, and the generation of metallic iron.

Metallic iron is more than twice as dense as any rock and if produced in any location in large enough quantities, it would sink downwards. If the production was only in droplets within a rock matrix it might still make a large body of such a material dense enough to force its way down. Whatever driving forces for motion we can invent in the mantle of the Earth, it will be hard to beat the one due to the production and sinking of iron towards the core.

An initially unmixed, uneven material of the forming Earth will be far from chemical equilibrium, and many chemical reactions could take place that give out energy and take it nearer to a final equilibrium. The chief bottleneck would be the fact that little motion is taking place, and the different chemicals that could react with each other do not meet. Any place where a large driving force for motion has once been created and where shearing motions therefore occur will in turn become a location where more chemical energy can be liberated. Perhaps a downgoing column of iron stirs up the surrounding medium, causing more chemical reactions and through that the production of more iron. In this way, spots or surfaces that are once activated may be expected to become even more active, while areas that are quiescent might remain so. The very uneven distribution of internal activity that we see in the form of the circum-Pacific belt and a few other active lines could have an explanation in this feedback mechanism.

There is another point to consider. One discusses the motion of large crustal plates like, for example, the whole North American continent, as reasonably coherent units. But, of course, a continent has no internal tensile strength at all. The crust is always broken, and tension in a rock is never maintained for any length of time. How then would it be possible to push a large plate over a big distance without it coming completely apart into minute pieces? Whatever the forces below might do, they cannot be expected to push everywhere in precisely parallel directions and yet, if there were the slightest divergence in these directions, the overlying crust would immediately come apart. In addition, a continental mass would, in any case, have internal forces in the sense of trying to spread out that mass and distribute it more widely. The fact that we have sharply defined continental blocks has to be explained first, and it surely means that there is some force acting inwards on each of them, holding them together. Any overall motion may then represent no more than a slight asymmetry of the forces pushing in from all sides on that continental plate.

An inward motion on some area could be expected if within that area a downward motion below is dominant, dragging the surface material towards the downgoing column. The crustal block, being made of a less dense material, refuses to be dragged down but merely tends to accumulate over this column. It is as with the scum in the sink which collects on the water above the drain-hole.

This speculation would then suggest that the thick continental material causes the mantle underneath to be somewhat hotter than it is under the much shallower oceanic crust, and this may tend to initiate the chemical reactions and the sinkage of iron.

Where the mantle is chemically active and where iron is sinking, rock material of smaller density will also be generated. This will tend to move upwards and add to the mass of the continental block. Even though both an upward and a downward motion would then be occurring in the same vicinity, we would still expect that the much more powerful downward motion of the sinking iron would overpower the large-scale pattern, and that the upwelling of the lighter rocks would be comparatively minor.

These, of course, as we have said, are speculations only. But when one considers that the deep Earth is still outgassing, that deep fluid channels exist, that volatiles are evolved and collected, that less dense materials, rock and fluids, are driven towards the surface, it is hard to avoid the conclusion that chemical processes are active and that denser materials are also produced. Those of course will sink away and we will not see them directly. Nevertheless, seismic and other investigations may allow us to build up some knowledge, where only speculation exists now.

How does this interact with the discussion of the origin of oil and gas? It is on the pattern of lines where the Earth is more active, or where it has been active in the past, that we tend to find more hydrocarbons. Although the chemical reduction of iron will tend to liberate carbon dioxide and water, any motion at these deeper levels may also mobilize other volatile substances and allow them to start their upward migration. The understanding of the mechanics of the deep Earth may one day help us to find the reasons for the particular distribution of oil and gas over the globe.

REFERENCES

Agnew, D. C. (1978), The 1952 Fort Yuma earthquake – two additional accounts. *Seismol. Soc. Amer. Bull.* **68**, 1761-1762.

Aguilera, J. G. (1920). The Sonora Earthquake of 1887. *Seismol. Soc. Amer. Bull.* **10**, 31-44.

Aldrich, L. T., and Nier, A. O. (1948). The occurrence of He^3 in natural sources of helium. *Phys. Rev.* **74**, 1590-1594.

Alippi, T. (1911). Nuovo contributo all'inchiesta sui "brontidi". *Boll. Soc. Sismol. Ital.* **15**, 65-77.

Ali-Zade, A. A., Shnyukov, E. F., Grigoryants, B. V., Aliyev, A. A., and Rakhmanov, R. R. (1984). Geotectonic conditions of mud volcano manifestation in the world and their role in prediction of gas and oil content in the Earth's interior. *Proc. 27th Internat. Geol. Congr.* **13**, 377-393.

Anders, E., Hayatsu, R., and Studier, M. H. (1973). Organic compounds in meteorites. *Science* **182**, 781-790.

Antesiforov, A. E., Petukhov, A. V., Trofinov, M. V., Fursov, V. Z., and Shpagin, D. E. (1983). Possibilities of mercury gas survey for oil and gas explorations. *Soviet Geology* (8), 108-113.

Arculus, R. J., and Delano, J. W. (1980). Implications for the primitive atmosphere of the oxidation state of the Earth's upper mantle. *Nature* **288**, 72-74.

Bagnold, T. (1829). Extraordinary Effect of an Earthquake at Lima, 1828. *Quart. J. Sci. Lit. Art* **27**, 429-430.

Barker, C., and Dickey, P. A. (1984). Hydrocarbon habitat in main producing areas, Saudi Arabia: discussion. *Amer. Assoc. Petrol. Geol. Bull.* **68**, 108-109.

Beskrovny, N. S. and Tikhomirov, N. I. (1968). Bitumens in the hydrothermal deposits of Transbaykal. In: The genesis of oil and gas; Izdvo Nedra.

Bischof, G. (1839). On the natural history of volcanoes and earthquakes. *Amer. J. Sci.* **37**, 41-77.

Buskirk, R. E., Frohlich, C., and Latham, G. V. (1981). Unusual animal behavior before earthquakes: a review of possible sensory mechanisms. *Rev. Geophys. Space Phys.* **19**, 247-270.

Chekaliuk, E. B. (1976). The thermal stability of hydrocarbon systems in geothermodynamic conditions. *Degazatsiia Zemli i Geotektonika* (P. N. Kropotkin, ed.), pp. 267-272.

Cox, K. G. (1978), Kimberlite pipes. *Scientific American* **238** (4).

Davis, T. B. (1984). Subsurface pressure profiles in gas-saturated basins. *Elmworth – Case Study of a Deep Basin Gas Field* (J. A. Masters, ed.), pp. 189-204. AAPG: Tulsa.

Dawson, R. F. (1981). Hydrogen production. *Chem. & Eng. News* 59 (15), 2.

Degens, E. T., von Herzan, R. P., How-Kin Wong, Deuser, W. G., and Jannasch, H. W. (1973). Lake Kivu: structure, chemistry and biology of an East African Rift lake. *Geol. Rundschau* 62, 245-277.

Demetrescu, G., and Petrescu, G. (1941). Sur les phénomènes lumineux qui ont accompagné le tremblement de terre de Roumanie de 10 Novembre 1940. *Acad. Roumaine Bull. Sec. Sci.* 23, 292-296.

Deng, Q., Jiang, P., Jones, L. M., and Molnar, P. (1982). A preliminary analysis of reported changes in ground water and anomalous animal behavior before the 4 February 1975 Haicheng earthquake. *Earthquake Prediction: An International Review* (D. W. Simpson and P. G. Richards, eds.), pp. 543-565. American Geophysical Union, Washington, D.C.

Deuser, W. G., Degens, E. T., Harvey, G. R., and Rubin, M. (1973). Methane in Lake Kivu: new data bearing on its origin. *Science* 181, 51-54.

Donovan, T. J. (1974). Petroleum microseepage at Cement, Oklahoma: evidence and mechanism: *Amer. Assoc. Petrol. Geol. Bull.* 58, 429-446.

Donovan, T. J., Forgey, R. L., and Roberts, A. A. (1979). Aeromagnetic detection of diagenetic magnetite over oil fields. *Amer. Assoc. Petrol. Geol. Bull.* 63, 245-248.

Donovan, T. J., Friedman, I., and Gleason, J. D., (1974). Recognition of petroleum bearing traps by unusual isotopic compositions of carbonate-cemented surface rocks. *Geology* 2, 351-354.

Duchscherer, W. Jr. (1981). Nongasometine geochemical prospecting for hydrocarbons with case histories: *Oil & Gas J.*, 19 Oct 1981, 312-327.

Duchscherer, W. Jr. (1984). *Geochemical Hydrocarbon Prospecting with Case Histories.* Penn Well Publishing Company: Tulsa, Okla.

Dyck, W. and Dunn, C. E. (1986). Helium and methane anomalies in Southwestern Saskatchewan, Canada, and their relationship to other dissolved constituents, oil and gas fields, and tectonic patterns. 91, B12, 12, 343-12, 353. *J. Geophys. Res.*

Fitz-Roy, R. (1836). Sketch of the surveying voyages of his majesty's ships Adventure and Beagle, 1825-1836. *J. Roy. Geogr. Soc.* 6, 311-343.

Fuller, M. L. (1912). The New Madrid Earthquake. *U.S. Geol. Surv. Bull.* 494.

Galimov, E. M. (1969). Isotopic composition of carbon in gases of the crust. *Internat. Geol. Rev.* 11, 1092-1103.

Galimov, E. M., and Kvendvolden, K. A. (1983). Concentrations and carbon isotopic compositions of CH_4 and CO_2 in gas from sediments of the Blake Outer Ridge, Deep Sea Drilling Project Leg 76. *Initial Rept. Deep Sea Drilling Project* 76, 403-407.

Galli, I. (1911). Raccolta e classificazione di fenomeni luminosi osservati nei terremoti. *Bol. Soc. Sismol. Ital.* 14, 221-447.

Gerlach, T. M. (1980). Chemical characteristics of the volcanic gases from Nyiragongo lava lake and the generation of CH_4-rich fluid inclusions in alkaline rocks. *J. Volcanol. Geotherm. Res.* 8, 177-189.

Gold, T. (1955). Geophysical discussion. *Observatory* 75, 110-113.

Gold, T. (1979). Terrestrial sources of carbon and earthquake outgassing. *J. Petrol. Geol.* 1 (3), 3-19.

Gold, T. (1982). Abiogenic derivation of petroleum: a reply to Treibs and Hirner. *Erdol & Kohl-Erdgas-Petrochemie* 35, 508-509.

Gold, T. (1984). Contributions to the theory of an abiogenic origin of methane and other terrestrial hydrocarbons. *Proc. 27th International Geological Congress, Vol. 13 (Oil and Gas Fields)*, pp. 413-442. VNU Science Press.

Gold, T. (1985). The origin of natural gas and petroleum, and the prognosis for future supplies. *Annual Review Energy* 10, 53-77.

Gold, T. (1986). Oil from the [depths] of the Earth. *New Scientist* 26, 42-46.

Gold, T. (1986). The origin of natural gas and petroleum. *Mendeleev Chemistry Journal*, special issue, in press.

Gold, T., and Held, M. (1987). Helium-nitrogen-methane systematics in natural gases of Texas and Kansas. *J. Geophys. Res.*, special issue, in press.

Gold, T., and Soter, S. (1979). Brontides: natural explosive noises. *Science* 204, 371-375.

Gold, T., and Soter, S. (1980). The deep Earth gas hypothesis. *Scientific American* 242 (6), 154-161.

Gold, T., and Soter, S. (1981). Natural explosive noises. *Science* 212, 1297-1298.

Gold, T., and Soter, S. (1982). Abiogenic methane and the origins of petroleum. *Energy Exploration & Exploitation* 1 (2), 89-104.

Gold, T., and Soter, S. (1983). Methane and seismicity: a reply. *EOS* 64, 522.

Gold, T. and Soter, S. (1985). Fluid ascent through the solid lithosphere and its relation to earthquakes. *Pure & Appl. Geophys.* 122, 492-530.

Gold, T., and Soter, S. (1986). Biogenic and abiogenic petroleum. *Chem. & Eng. News* 64 (16), 2-3.

Gold, T., Gordon, B. E., Streett, W. E., Bilson, E. and Patnaik, P. (1986). Experimental study of the reaction of methane with petroleum hydrocarbons under geological conditions. *Geochim. Cosmochim. Acta* 50 (11), in press.

Golubev, O. A., Kolobashkin, V. M., Popov, A. I., Popov, E. A., and Fabritsius, Z. E. (1984). On the relationship between the methane content of near-surface air and current geodynamic processes. *Volc. Seis.* 5, 223-226.

Goodfellow, G. E. (1888). The Sonora Earthquake. *Science* 11, 162-166.

Hauksson, E. (1981). Radon content of groundwater as an earthquake precursor: evaluation of worldwide data and physical basis. *J. Geophys. Res.* 86. 9397-9410.

Haywood, J. (1823). *The Natural and Aboriginal History of Tennessee.* Nashville, Tenn.

Hedberg, H. D. (1964). Geologic aspects of origin of petroleum. *Amer. Assoc. Petrol. Geol. Bull.* 48, 1755-1803.

Henderson, G. (1964). Oil and gas prospects in the Cretaceous-Tertiary basin of West Greenland. *Geol. Survey Greenland Rept. No. 22.*

Hentig, H. von (1923). Animal and Earthquake. *J. Compar. Psychology* 3, 61-71.

Higgins, G. E., and Saunders, J. B. (1974). Mud volcanoes – their nature and origin. *Verhandl. Naturf. Ges. Basel* 84, 101-152.

Hitchcock, C. H. (1865). The Albert coal, or Albertite of New Brunswick. *Amer. J. Sci., 2nd Ser.* 39, 267-273.

Hodgson, G. W., and Baker, B. L. (1969). Porphyrins in meteorites: metal complexes in Orgueil, Murray, Cold Bokkeveld, and Mokoia-carbonaceous chondrites. *Geochim. Cosmochim. Acta* 33, 943-958.

Hoefs, J. (1972). Is biogenic carbon always isotopically "light", is isotopically "light" carbon always of biogenic origin? *Adv. Organic Geochem. 1971*, pp. 657-663.

Holmes, A. (1944-1955). *Principles of Physical Geology.* Thomas Nelson and Sons Ltd: Edinburgh.

Hovland, M., Judd, A. G., and King, L. H. (1984). Characteristic features of pockmarks on the North Sea floor and Scotian Shelf. *Sedimentology* 31, 471-480.

Hoyle, F. (1955). *Frontiers of Astronomy.* Harper and Brothers: New York.

Humboldt, A., von (1822). *Personal Narrative of Travels to the Equinoctial Regions of the New Continent during the Years 1799-1802*, Vol.2 (London, 3rd ed.), pp. 211-212.

Ippolito, F. (1793). An Account of the Earthquake which Happened in Calabria, March 28, 1783. *Phil. Trans. Roy. Soc.* 73, Appendix i-vii.

Jones, P. H. (1980). Role of geopressure in the hydrocarbon and water system. *Problems of Petroleum Migration* (W. H. Roberts III and R. J. Cordell, eds.), pp. 207-216. Amer. Assoc. Petrol. Geol., Tulsa.

Kanamori, H., Ekstrom, G., Dziewonski, A., Barker, J. S. (1986). An anomalous seismic event near Tori Shima, Japan: a possible magma injection event. *EOS* 67, 1117.

Kapo, G. (1978). Vanadium: key to Venezuelan fossil hydrocarbons. *Bitumens, Asphalts and Tar Sands* (G. V. Chilingarian and T. F. Yen, eds.), pp. 213-241. Elsevier: Amsterdam.

Kennedy, G. C., and Nordlie, B. E. (1968). The genesis of diamond deposits. *Econ. Geol.* 63, 495-503.

Kent, P. E., and Warman, H. R. (1972). An environmental review of the world's richest oil-bearing region – the Middle East. *Internat. Geol. Congr. 24th, Sect. 5*, pp. 142-152.

Kim, K., Craig, H., and Horibe, Y. (1983). Methane: a "real-time" tracer for submarine hydrothermal systems. *EOS* 64, 724.

Kozlovsky, Ye. A. (1984). The world's deepest well. *Scientific American* 251 (6), 98-104.

Kravtsov, A. I. (1975). Inorganic generation of oil and criteria for exploration for oil and gas. *Zakonomern. Obraz. Razmeshchniya Prom. Mestorozhd. Nefti Gaza* (G. N. Dolenko, ed.), pp. 38-48. Naukova Dumka: Kiev.

Kravtsov, A. I., Ivanov, V. A., Bobrov, V. A., and Kropotova, O. I. (1981). Distribution of gas-oil-bitumen shows in the Yakutian diamond province. *Internat. Geol. Rev.* 23, 1179-1182.

Kravtsov, A. I., Voytov, G. I., Ivanov, V. A., and Kropotova, O. I. (1976). Gases and bitumens in rocks of the Udachnaya pipe. *Dokl. Akad. Nauk SSSR, Earth Sci. Sect.* 228, 231-234.

Kropotkin, P. N. (1985). Degassing of the Earth and the origin of hydrocarbons. *Internat. Geol. Rev.* 27, 1261-1275.

Kropotkin, P. N., and Valaiev, B. M. (1976). Development of a theory of deep-seated (inorganic and mixed) origin of hydrocarbons. *Goryuchie Iskopaemye: Problemy Geologii i Geokhimii Naftidov i Bituminoznykh Porod* (N. B. Vassoevich, ed.), pp. 133-144. Akad. Nauk SSSR.

Kropotkin, P. N., and Valyaev, B. M. (1984). Tectonic control of Earth outgassing and the origin of hydrocarbons. *Proc. 27th Internat. Geol. Congr.* 13, 395-412. VNU Science Press.

Kudryavtsev, N. A. (1959). Geological proof of the deep origin of petroleum. *Trudy Vsesoyuz. Neftyan. Nauch. -Issledovatel. Geologoraz Vedoch. Inst.* No. 132, 242-262.

Kvenvolden, K. A., and Barnard, L. A. (1982). Hydrates of natural gas in continental margins. *Studies in Continental Margin Geology* (J. S. Watkins and C. L. Drake eds.), pp. 631-640. AAPG Memoir 34.

Lacroix, A. (1904). *La Montagne Pelée et ses Eruptions*, pp. 194-195.

Lalemant, H. (1663). Relation of What Occurred Most Remarkable in the Missions of the Fathers of the Society of Jesus in New France in the Years 1662 and 1663. *The Jesuit Relations and the Allied Documents* (R. G. Thwaites, ed.), Vol. 48, pp. 17-179.

Landes, K. K. (1970). *Petroleum Geology of the United States*, pp. 106-118. John Wiley & Sons, Inc.: New York.

Larkin, E. L. (1906). The great San Francisco earthquake. *Open Court* 20, 393-406.

Lawson, A. C., et al. (1908). *The California Earthquake of April 18, 1906.* Carnegie Institution, Washington, D.C.

Lee, W. H. K., Ando, M., and Kautz, W. H. (1976). A summary of the literature on unusual animal behavior before earthquakes. *U.S. Geol. Survey Open-file Rept.* 76-826.

Le Guern, F., Tazieff, H., and Fairre Pierret, R. (1982). An example of health hazard: people killed by gas during a phreatic eruption: Diëng Plateau (Java, Indonesia), February 20th, 1979. *Bull. Volcanol.* 45 (2), 153-156.

Li, J. (1980). Earthquake – a harvest of agony. *Los Angeles Times*, Oct. , 1980.

Liao-ling Province Meteorological Station (1977). The extraordinary phenomena in weather observed before the February 1975 Hai-cheng earthquake. *Acta Geophys. Sinica* 20, 270-275.

Lupton, J. E. (1983). Terrestrial inert gases: isotope tracer studies and clues to primordial components in the mantle. *Ann. Rev. Earth Planet Sci.* 11, 371-414.

Mallet, R. (1852, 1853, 1854). Third Report on the Facts of Earthquake Phenomena. *Report Brit. Assoc. Adv. Sci.* 22, 1-176; 23, 117-212; 24, 1-326.

Melton, C. E., and Giardini, A. A. (1974). The composition and significance of gas released from natural diamonds from Africa and Brazil. *Amer. Mineralogist* 59, 775-782.

Mendeleev, D. (1877). L'origine du petrole. *Revue Scientifique, 2e Ser.*, VIII, 409-416.

Michell, J. (1761). Conjectures concerning the cause, and observations upon the Phaenomena, of Earthquakes. *Phil. Trans. Roy. Soc.* 51, 566-634.

Miller, R. D., and Hertwick, F. R., Jr. (1982). Analyses of Natural Gases, 1981 *U.S. Bur. Mines Info. Circ. 8890.*

Miller, R. D., and Hertwick, F. R., Jr. (1983). Analyses of Natural Gases, 1982. *U.S. Bur. Mines Info. Circ. 8942.*

Milne, J. (1986). *Earthquakes and other earth movements.* 1st ed. D. Appleton & Co., New York; 6th ed. (1913) Kegan, Paul, Trench, Trubner & Co., London.

Moore, B. J. (1982). Analyses of Natural Gases, 1917-80. *U.S. Bur. Mines Info. Circ. 8870.*

Newton, I. (1730). *Optics*, 4th ed., q. 31, pp. 354-355.

Nikonov, V. F. (1973). Formation of helium-bearing gases and trends in prospecting for them. *Internat. Geol. Rev.* 15, 534-541.

Ninkovich, D., Sparks, R. S. J., and Ledbetter, M. T. (1978). The exceptional magnitude and intensity of the Toba eruption, Sumatra: an example of the use of deep-sea tephra layers as a geological tool. *Bull. Volcanol.* 41, 286-298.

Ourisson, G., Albrecht, P., and Rohmer, M. (1984). The microbial origin of fossil fuels. *Scientific American* 251 (2), 44-51.

Pasteris. J. D. (1983), Kimberlites: a look into the Earth's mantle. *Amer. Scientist* 71, 282-288.

Pedersen, K. R., and Lam, J. (1970). Precambrian organic compounds from the Ketilidian of South-West Greenland. *Gronlands Geologiske Unders. Bull. No. 82.*

Philippi, G. T. (1977). On the depth, time and mechanism of origin of the heavy to medium-gravity napthenic crude oils. *Geochim. Cosmochim. Acta* 41, 33-52.

Pierce, A. P., Gott, G. B., and Mytton. J. W. (1964). Uranium and helium in the Panhandle Gas Field, Texas, and adjacent areas. *U.S. Geol. Survey Prof. Paper 454-G*.

Pierce, W. L. (1812). *Evening Post*, New York, 11 February 1812.

Pogorski, L. A., and Quirt, G. S. (1979). Helium emanometry in exploring for hydrocarbons, I. *Unconventional Techniques in Exploration for Petroleum and Natural Gas*, pp. 124-129. Southern Methodist Univ. Press: Dallas, Texas.

Porfir'ev, V. B. (1974). Inorganic origin of petroleum. *Amer. Assoc. Petrol. Geol. Bull.* 58, 3-33.

Powers, S., *et al.* (1932). Symposium on the occurrence of petroleum in igneous and metamorphic rocks. *Amer. Assoc. Petrol. Geol. Bull.* 16 (8).

Pratt, W. E. (1952). Towards a philosophy of oil-finding. *Bull. AAPG* 36, 12, pp. 2231-1136.

Price, L. C. (1982). Organic geochemistry of core samples from an ultradeep hot well (300°C, 7 km). *Chem. Geol.* 37, 215-228.

Raffles, T. S. (1817). *The History of Java*, Vol. 1, p. 28.

Reitsema, R. H. (1979). Gases of mud volcanoes in the Copper River Basin, Alaska. *Geochim. Cosmochim. Acta* 43, 183-187.

Rethly, A. (1952). *A Kárpátmedencék Földrengesei 445-1918*. Academic Publishing House: Budapest.

Rikitaki, T. (1976). *Earthquake Prediction*. Elsevier: New York.

Roberts, A. A. (1979). Helium emanometry in exploring for hydrocarbons, II. *Unconventional Techniques in Exploration for Petroleum and Natural Gas*, pp. 136-149. Southern Methodist Univ. Press: Dallas, Texas.

Robinson, R. (1963). Duplex origin of petroleum. *Nature* 199, 113-114.

Robinson, R. (1966). The origins of petroleum. *Nature* 212, 1291-1295.

Ronov, A. B., and Yaroskevskiy, A. A. (1976). A new model for the chemical structure of the Earth's crust. *Geochem. Internat.* 13 (6), 89-121.

Rubey, W. W. (1951). Geologic history of sea water – an attempt to state the problem. *Geol. Soc. Am. Bull.* 62, 1111-1147.

Rudakov, G. V. (1973). The relationship between incidences of mercury mineralization and the presence of oil and gas. *Geolgichnii Zh.* 33 (5), 125-26.

Schidlowski, M. (1982). Content and isotopic composition of reduced carbon in sediments. *Mineral Deposits and the Evolution of the Biosphere* (H. D. Holland and M. Schidlowski, eds.), pp. 103-122. Springer-Verlag: Berlin.

Schidlowski, M., Hayes, J. M., and Kaplan, I. R. (1983). Isotopic inferences of ancient biochemistries: carbon, sulfur, hydrogen, and nitrogen.

Earth's Earliest Biosphere (J. W. Schopf, ed.), pp. 149-186. Princeton Univ. Press.

Schmidt, A., and Mack, K. (1913). Das Süddeutesches Erdbeben vom 16 November 1911. *Württ, Jahrbücher f. Statist. u. Landeskde., Jahrg. 1912, Heft I*, 96-139.

Simon, C. (1663). An Account of the Earthquake in New France, 1663. *The Jesuit Relations and Allied Documents* (R. G. Thwaites, ed.), Vol. 48, pp. 181-223.

Smith, R. I. L., and Clymo, R. J. (1984). An extraordinary peat-forming community on the Falkland Islands. *Nature* 309, 617-620.

Sokoloff, W. (1889). Kosmischer Ursprung der Bitumina. *Bull. Soc. Imp. Natural Moscau, Nouv. Ser.* 3, 720-739.

Sokolov, V. A., Buniat-Zade, Z. A., Goedekian, A. A., and Dadashev, F. G. (1963). The origin of gases of mud volcanoes and the regularities of their powerful eruptions. *Adv. Organic Geochem. 1968* (P. A. Schenck and I. Havenaar, eds.), pp. 473-484. Pergamon Press.

Stehn, C. E. (1929). The geology and volcanism of the Krakatau group. *Proc. Fourth Pacific Science Congress (Batavia)*, pp. 1-55. [Reprinted in T. Simkin and R. Fisk (1983), *Krakatau 1883*, Smithsonian Inst. Press, Washington.]

Stoqueler, Mr. (1756). Observations, Made at Colares, on the Earthquake at Lisbon, of the 1st of November 1755, by Mr. Stoqueler, Consul of Hamburg. *Phil Trans. Roy. Soc.* 49, 413-418.

Sugisaki, R. (1978). He/Ar and N_2/Ar ratios of fault air may be earthquake precursors. *Nature* 275, 209-211.

Sugisaki, R., and Sugiura, T. (1985). Geochemical indicator of tectonic stress resulting in an earthquake in central Japan, 1984. *Science* 229, 1261-1262.

Sylvester-Bradley, P. C. (1964). The origin of oil and life. *Discovery* 25, 37-42.

Sylvester-Bradley, P. C. (1971). Environmental parameters for the origin of life. *Proc. Geologists' Assoc.* 82, 87-135.

Sylvester-Bradley, P. C. (1972). The geology of juvenile carbon. *Exobiology* (C. Ponnamperuma, ed.), pp. 62-94.

Tietze, K., Geyh, M., Muller, H., Schroder, L., Stahl, W., and Wehner, H. (1980). The genesis of methane in Lake Kivu (Central Africa). *Geol. Rundschau* 69, 452-472.

Valyayev, B. M. and Grinchenko, Yu. I. (1985). The Origin of Isotopically Ultraheavy Carbon Dioxide. *Internat. Geol. Rev.* 27:11, pp. 1315-1324.

Vernadksky, V. I. (1933). The history of minerals of the Earth's crust, Vol. 2, Pt. 1 (in Russian). Moscow-Leningrad.

Vlasov, K. A., editor (1968). *Geochemistry and Mineralogy of Rare Elements and Genetic Types of Their Deposits. Volume III: Genetic Types of Rare-Element Deposits*, pp. 719-737. Israel Program for Scientific Translation: Jerusalem.

Vornoi, E. E. (1984). The hydrogenation mobilization of kerogen-coal in rock by deep hydrocarbons. *Proiskhozhdemi i Migratsiia Nefti i Gaza* (G. N. Dolenko *et al.*, eds.), pp. 111-119.

Wakita, H., and Sano, Y. (1983). $^3He/^4He$ ratios in CH_4-rich natural gases suggest magmatic origin. *Nature* 305, 792-794.

Wallace, R. E., and Ta-Liang Teng. (1980). Prediction of the Sungpan-Pingwu Earthquakes, August 1976. *Amer. Seismol. Soc. Bull.* 70, 1199-1223.

White, B. (1985). Frozen energy in the depths of the oceans. *New Scientist*, 4 July 1985, p. 29.

Wilkening, L. (1978). Carbonaceous chondritic material in the solar system. *Naturwiss.* 65, 73-79.

Williams, S. (1785). Observations and Conjectures on the Earthquakes of New-England. *Amer. Acad. Arts & Sci. Memoirs* 1, 260-311.

Winslow, C. F. (1865). Earthquake at Lima, March, 1865. *Amer. J. Sci.*, 2nd Ser. 40, 365.

Wittich, W. (1869). *Curiosities of Physical Geography*, pp. 280-281.

Note: there are variations in the English spelling of some of the Soviet authors.

GLOSSARY

Abiogenic: Not derived from biological materials.

Alluvium: A sediment which is not hardened into a rock. Usually water-transported grains such as clay.

Archean: The oldest era of the geologic record. More than 2,500 million years ago.

Argon: Element No. 18. A noble gas present in the atmosphere at a level of approximately 1 per cent. Argon has three stable isotopes, argon 40, argon 36 and argon 38. Terrestrial argon is predominantly argon 40, produced by the radioactive decay of potassium. Argon 36 and 38 are present in small abundance and are predominantly of primordial origin.

Astrobleme: An area on the Earth where a large meteorite struck leaving a crater or an area of shattered or modified rock.

Basin (sedimentary basin):   A depression of the basement rock filled in by sediments (example: Anadarko Basin in Oklahoma. The deepest part is approximately 45,000 feet).

Biogenic: Substances derived from biological materials.

Bitumen: Dense, very viscous hydrocarbon compounds, somewhat oxidized as for example tar and asphalt. In earlier times the term was used to include all natural petroleum.

Brontides: Natural air-borne noises generated by some effect originating in the ground; frequently loud explosive noises, sometimes associated with earthquakes and sometimes apparently precursors to earthquakes. It is thought that the origin is frequently the sudden emission of gas from the ground.

Cap rocks: Dense rocks that obstruct or delay the upward seepage of hydrocarbon fluids. Layers of salt and clay often form cap rocks, as they are dense and not so likely to develop fractures as other more brittle rocks.

Carbon: Element number 6; it has two stable isotopes, carbon-12, six protons, six neutrons, and carbon-13, six protons, seven neutrons. Natural terrestrial carbon is approximately 99% carbon-12, 1% carbon-13. Carbon-14, six protons and eight neutrons, is radioactive with a half-life of 8600 years. It is created in the atmosphere from nitrogen by the action of cosmic rays, and is only present in surface materials in trace amounts.

Carbon dioxide: CO_2, a constituent of the atmosphere at a level of 0.03% by volume. Atmospheric carbon dioxide appears to have been the chief source of carbon in the carbonate rocks, and it is the chief source of the carbon in plants.

Carrier gas: Any gas that establishes a bulk flow through permeable rocks and thereby picks up and transports other substances which, by themselves, would have remained immobile.

Cosmochemistry: The chemistry of the tenuous media in space, including that of the clouds that surrounded the sun and were responsible for the formation of the planetary system.

Cretaceous: Geologic period in the Mesozoic era from 136 to 65 million years ago approximately.

Crust (of the Earth): Outermost layer of rock, 10–60km thick, less dense, harder and more brittle than the (mantle) material below. Derived apparently mainly from partial melting and extrusion from the mantle.

Crystalline basement: The rocks underneath the sediments which give evidence of having been emplaced at a sufficiently high temperature so that on cooling they developed a structure of intertwined crystals.

Diamonds: The crystal configuration of pure carbon, metastable at low pressures, but the stable (equilibrium) configuration at pressures like those that occur in the Earth below about 150km depth.

Diffusion: The transport of molecules through a substance when the only driving force is random thermal motion.

Fischer-Tropsch Process: A process whereby liquid hydrocarbons are produced in chemical reactions involving carbon monoxide and hydrogen. A process that has been used on a commercial scale to produce liquid fuels when petroleum products were not available.

Gneiss: A major type of crustal rock, metamorphic, meaning that it has been exposed to heat at least sufficient to allow it to recrystallise.

Granite: A type of rock which is a major component of the continental crust. It is a metamorphic rock having been heated and cooled sufficiently to form crystals. Details of the origin of granite are not clear and are still under debate.

Helium: Element No. 2. A noble gas, the second most abundant element in the universe and in the sun, but present in very small amounts only on the Earth. Helium has two stable isotopes, helium 4, two protons, two neutrons, and helium 3, two protons, one neutron. Helium 4 is created in radioactive decay processes mainly of uranium and thorium, and the nucleus of helium 4 is then referred to as an alpha particle. Most terrestrial helium was produced in radioactive decay processes and only a small fraction was supplied in the process of formation of the Earth. That fraction referred to as primordial helium, is characterised by possessing a higher proportion of helium 3.

Hydrocarbons: Compounds composed mainly of carbon and hydrogen, although small amounts of nitrogen, oxygen and other elements may be included also in the molecular structure. Methane, CH_4, is the lightest. Petroleum is composed of a large range of hydrocarbon molecules of different structural configurations, of different hydrogen/carbon ratios and of a great range of molecular weights. Kerogen and coal are also classified as hydrocarbons.

Igneous rock: A rock frozen from a melt.

Jurassic: Geologic period of the Mesozoic era from about 195 TO 136 million years ago.

Kudryavtsev's Rule: The tendency, widely identified, that underneath any oil field some petroleum can be found at all deeper levels down to and sometimes into the crystalline basement.

Magma: Any hot, liquid rock under the surface.

Mantle (of the Earth): The region of the body of the Earth above the core and below the crust. The radius of the liquid core is approximately half the radius of the Earth and the crust is a thin exterior layer between 10 and 60 km thick. Approximately 7/8ths of the volume of the Earth is therefore made up by the mantle, composed of a dense solid at a temperature at which it is subject to plastic deformation and partial melting. At least the outer mantle is known to be of regionally uneven chemical composition.

Metamorphic rocks: Rocks that give evidence of having been substantially modified by heat.

Meteorites: Stony or metallic objects from space that strike the surface of the Earth (distinguished from meteors which are objects too small to penetrate the atmosphere without burning up). The meteorites are thought to be debris left over from the period of formation of the planets. They therefore give evidence of the types of material and the chemical condition responsible for planetary formation.

Methane: CH_4, the lightest and most volatile of the hydrocarbons, and the principal constituent of natural gas.

Methane hydrates: An ice composed mainly of water together with a few per cent methane. The crystal structure is called a "clathrate" in which the methane molecules fit into the water ice crystal structure. Under pressure this ice freezes at substantially higher temperatures than plain water ice and can therefore exist on or below the ocean floor as well as at depths in permafrost regions where water ice would not be stable.

Miocene: A geologic epoch in the tertiary period from about 22 to 5 million years ago.

Mud volcanoes: Mountains or mounds from which subterranean mud is periodically expelled, creating a feature resembling a volcanic cone with a crater on top.

Oligocene: A geologic epoch in the tertiary period from about 37 to 22 million years ago.

Paleozoic: The geologic era dating from approximately 570 to 225 million years ago.

Permeability: The ability of the ground to permit the flow and transport of fluids through it. Interconnecting pore spaces or cracks are a requirement for permeability to exist.

Petroleum: All naturally occurring liquid hydrocarbons. Literally "stone oil".

Photosynthesis: The process whereby plants absorb carbon from atmospheric carbon dioxide turning it into unoxidised carbon. The energy for this process is supplied by sunlight. In the same action water is also dissociated and oxygen is liberated into the atmosphere.

Pockmarks: A term used to describe circular markings that have been observed on many areas of the ocean floor. They are thought to arise from sudden gas emissions, with the consequent stirring up and resettling of the ocean mud.

Porosity: The existence of pore spaces in rocks or sediments. It is usually described by the proportion of volume occupied by the pores. Pore spaces in the ground are filled with fluids such as air, water, natural gas, petroleum, carbon dioxide, nitrogen, hydrogen and smaller amounts of other gases or liquids.

Pre-Cambrian: The geologic era pre-dating the Cambrian period which commenced approximately 570 million years ago.

Radon: Element number 86. A noble gas but with no stable isotopes. Radon 222 is created in the radioactive decay and it has a half-life of 3.8 days. Because of its short half-life it is useful as an indicator for the flow of gases through rocks, in which traces of this gas are constantly produced by the uranium that is present.

Reservoir rocks: Porous rocks that contain hydrocarbon fluids.

Silurian: A geologic period in the Paleozoic era between 435 and 395 million years ago. It is the period in which land vegetation developed and became widespread.

Source rock: A rock whose content of unoxidized carbon compounds was considered to have provided the petroleum or natural gas of productive fields in the vicinity. On the basis of the theories of a biological origin of terrestrial hydrocarbons such a source rock of adequate content to have supplied these fields had to exist, and its carbonaceous material was considered to be the remains of plant or animal debris. On the basis of an abiogenic origin of most terrestrial hydrocarbons, oil and gas fields may exist without identifiable source rocks in the vicinity. Where such carbonaceous material is found it may be largely of non-biological origin and it may derive from the same deep-seated source that supplied the oil or gas of the area.

Tsunamis: Sometimes referred to as tidal waves. Ocean waves, usually of long period, caused by earthquakes, volcanic eruptions or large landslides. Tsunamis are sometimes devastating by suddenly flooding large areas of low lying land.

Volatiles: Gaseous substances or substances that can readily form gases.

INDEX

Achaia 50
Acid rain 173
Adriatic 73
Aegean 73
Aelian 51, 66
Agnew, D. C. 59
Alaska 150, 158, 162
Albert, Lake 129
Albertite 159
Albrecht, P. (with Ourisson) 143, 156
Aliyev, A. A. (with Ali-Zade) 100
Ali-Zade, A. A. 100
Anadarko Basin 5, 133, 134, 158
Anaxagoras 49
Andaman Islands 128
Anders, E. 18, 34, 35
Antesiforov, A. E. 148
Appalachian Mountains 158
Archaean 182
Arcs of circles 35
Arctic Ocean 162
Argon 22, 78, 121
Arica earthquake (1868) 59
Aristotle 45, 49
Arkansas 47
Asteroids 27
Athabasca tar sands 18
Azerbaizahn 181

Bagnold, T. 58
Baikal, Lake 17, 131
Baltic 73
Barker, C. 125
Barnard, L. A. (with Kvenvolden) 162
Bering Sea 73
Beskrovnyy, 18
Bilson, E. (with Gold) 136
Biological markers 142 ff

Bischof, G. 54
Bobrov, V. A. (with Kravtsov) 43
Bura 50
Burma 128
Buskirk, R. E. 70
Byron 93

Calabria (1783) 57
Calcite cements 152, 153
Calcium carbonate 151
Calderas 97
California, Gulf of 110, 128
Callisthenes 50
Canada 153, 162
Caprock 168
Carbon dioxide 24, 173
Carbon Isotopes 175
Carbonaceous chondrites 30, 31, 32, 34
Carbonaceous shale 153, 160
Carbonate rocks 23, 24
Carbonates 29
Chalcis 50
Charlevoix 47
Chekaliuk, E. B. 14, 138
China 128
Chu Chieh Cho 62
Cinnabar 148
Clymo, R. J. (with Smith) 150
Coal 153
Cobalt 148
Colombia 158
Comets 22, 31
Copper 148
Corinth, Gulf of 73
Craig, H. (with Kim) 110
Critical layers 168
Critical level 81
Cumana (Venezuela) earthquake (1797) 57

Davis, T. B. 91
Deep earthquakes 73
Degens, E. T. (with Deuser) 129
Delos 50
Demetrescu, G. 60
Deng, Q. 61, 69
Deuser, W. G. 129
Diamonds 14, 37, 38
Dickey, P. A. (with Barker) 125
Diffusion 78
Dolomite 152
Donetz Basin of the Ukraine 158
Donovan, T. J. 152, 170
Drozd, R. J. 71
Duchscherer, W. Jr. 170
Dunn, C. E. (with Dyck) 112
Dyck, W. 112

Earthquake lights 55, 64
Earthquake prediction 73
Earthquake spots 47
Earthquakes 45
East Pacific Rise 104, 128
Egmont, Mount 132
Elastic rebound 73, 77
Electrical conductivity 72
Enola 47
Eshelman, V. 27
Etna, Mount 16, 132

Fabritsius, Z. E. (with Golubev) 71
Falkland Islands 150
Fischer-Tropsch 150
Fischer-Tropsch process 34, 35
Fitz-Roy, R. 59
Flames (from volcanoes) 98
Flathead Lake 47
Fossils in coal 154
Fractionation 175
Friedman, I. (with Donovan) 152,
  170
Frohlich, C. (with Buskirk) 70
Fuller, M. L. 58

Fursov, V. Z. (with Antesiforov)
  148

Galimov's Rule 178
Galli, I. 55, 56, 57
Gallium 159
Gas Research Institute 172
Germane 159
Germanium 159
Geyh, M. (with Tietze) 130
Giardini, A. A. (with Melton) 42
Gleason, J. D. (with Donovan) 152,
  170
Gold, T. 7, 91, 136
Golubev, O. A. 71
Goodfellow, G. E. 60
Gordon, B. E. (with Gold) 136
Gott, G. B. (with Pierce) 112
Great Lakes 73
Green Tuff 110
Greenland 159
Grigoryants, B. V. (with Ali-Zade)
  100
Grinchenko 181
Groningen 128
Guatemala 162
Gulf States 124

Haicheng (China) earthquake
  (1975) 61, 68
Halley's comet 31
Halo patterns 168
Harvey, G. R. (with Deuser) 129
Hauksson, E. 71
Hayatsu, R. (with Anders) 18, 34,
  35
Haywood, J. 58
Hedberg, H. D. 9
Helike 50, 51, 66
Helium 11, 22, 78, 104
Henderson 159
Hentig von 67
Higgins, G. E. 100

Hitchcock, C. H. 160
Hoefs, J. 153
Hokkaido 157, 170
Holmes, Arthur 157
Hopanoid 143, 156
Horibe, Y. (with Kim) 110
Houser, J. E. 64
Hovland, M. 73
Hoyle, F. 7
Hugoton fields 113ff., 133
Hugoton Panhandle fields 133
Humboldt von, Alexander 55, 58, 68
Hydrogen 22, 130

Imperial Valley 128
Indonesia 157
Ippolito, F. 57
Iran 124, 158
Iraq 124
Iridium 32
Ivanov, V. A. (with Kravtsov) 43

Jamaican earthquake (1688) 53
Japan 169
Jiang, P. (with Deng) 61, 69
Jones, M. L. (with Deng) 61, 69
Jones, V. T. 71
Judd, A. G. (with Hovland) 73
Jupiter 22, 27

Kamchatka Peninsula 131
Kanamori, H. 77
Kansas 113ff., 133
Kanto earthquake (1923) 69
Kapo, G. 148
Kennedy, G. C. 39
Kent, P. E. 126
Kerch Peninsula 17
Kerogen 12, 25, 153, 161
Kim, K. 110
Kimberley 39
Kimberlite 39
Kimberlite pipes 39

King. L. H. (with Hovland) 73
Kivu, Lake 110, 129
Knudsen, J. M. 186
Kobes, Charles 63
Kola Peninsula 18, 169
Kolobashkin, V. M. (with Golubev) 71
Komarom (Hungary) earthquake (1763) 57
Kozlovsky, Ye.A. 169
Krakatau eruption (1883) 92
Krakatau eruption (1928) 98
Kravtsov, A. I. 18, 43, 131
Kropotkin, P. N. 17, 19, 20
Kropotova, O. I. (with Kravtsov) 43
Krypton 22
Kudryavtsev, N. A. 16, 17, 20
Kudryavtsev's rule 17, 20, 132, 178
Kurile Islands 18, 131, 170
Kvenvolden, K. A. 162

La Malbaie 48
Lam, J. (with Pedersen) 159
Landes, K. K. 133
Larkin, E. L. 63, 64, 65
Latham, G. V. (with Buskirk) 70
Lava 96
Lawson, A. C. 62, 65
Levin, 18
Lignite 149
Lima earthquake (1828) 58
Limestone 151
Lindbo, Tord 171
Lisbon earthquake (1755) 53, 56
Lost Soldier oil field 16
Lupton, J. E. 104

Mack, K. (with Schmidt) 60
Madeira, George 64, 65
Magellan 149
Magellan, Straits of 149
Malawi, Lake 129
Mallet, Robert 55

Mantle 33
Mariana Islands 131
Mars 22, 27, 28
Melton, C. E. 42
Mendeleev, D. 3, 9, 16, 19
Merapi eruption in Sumatra (1932) 17
Mercury 22, 30, 148, 159
Meteorites 29, 31
Methane hydrates 12, 162
Mexico, Gulf of, 73
Michell, John 52
Middle East 124
Milne, John 55
Molnar, P. (with Deng) 61, 69
Molybdenum 148
Monoun, Lake (Cameroons) eruption (1984) 99
Montana 47
Mud volcanoes 87, 99ff., 165
Muller, H. (with Tietze) 130
Mytton, J. W. (with Pierce) 112

Neon 22
Neptune 22, 27
New Madrid earthquake (1811–12) 58, 71
New Zealand 132
Newton, Sir Isaac 51
Nickel 143, 148
Nikonov, V. F. 111
Nitrogen 113
Norcia and Aquila (Italy) earthquake (1703) 56
Nordlie, B. E. 39
North Sea Trench 128
Norway 128
Norwegian trench 171
Nyiragongo, volcano 130
Nyos, Lake (Cameroons) eruption (1986) 99

Okinawa earthquake (1984) 77
Oklahoma 133

Optical activity 10, 142, 144
Organometallic complexes 147
Origin of life 20
Orinoco, Delta of 73
Osmium 32
Ourisson, G. 143, 156
Owens Valley earthquake (1872) 59
Oxygen fugacity 15

Paleozoic 133
Panhandle fields 113ff
Patnaik, P. (with Gold) 136
Pausanias 50
Peat 149, 150
Pedersen, K. R. 159
Pee Dee Belemnite 175
Pelee, Mont, eruption (1902) 98, 99
Pennsylvania 158
Peridotite 39
Permafrost 162
Persian Gulf 127, 133
Peru 162
Petrescu, G. (with Demetrescu) 60
Petukhov, A. V. (with Antesiforov) 148
Pierce, A. P. 58, 112
Platinum 32
Pliny 50
Pockmarks 72
Pogorski, L. A. 112
Pompeii 51
Popov, A. I. (with Golubev) 71
Porfir'ev, V. B. 18
Powers, S. 13
Precambrian 133
Price, L. C. 141
Progressive fractionation 178, 179

Quebec earthquake (1663) 56
Quirt, G. S. (with Pogorski) 112

Radon emission 74

Radon monitoring 71
Rakhmanov, R. R. (with Ali-Zade) 100
Red Sea 110, 127, 129
Rethly, A. 57
Rhodium 32
Rikitaki, T. 69
Roberts, A. A. 112
Robinson, R. 9, 19, 142
Rohmer, M. (with Ourisson) 143, 156
Ronov, A. B. 152
Rubey, W. W. 24
Rubin, M. (with Deuser) 129
Rudakov, G. V. 148
Rumania earthquake (1940) 60

San Francisco earthquake (1906) 62
San Juan Basin of New Mexico 133, 158
Sano, Y. (with Wakita) 110
Santa Rosa (1906) 62
Santorini eruption (3,500 years ago) 93
Santorini eruption (1866) 98
Satellites 31
Saturn 22, 27
Saudi Arabia 124, 158
Saunders, J. B. (with Higgins) 100
Scandinavia 153, 171
Schmidt, A. 60
Schroder, L. (with Tietze) 130
Scotian Shelf 73
Seneca 50, 51
Shnyukov, E. F. (with Ali-Zade) 100
Shpagin, D. E. (with Antesiforov) 148
Siberia 153, 162
Sicily 133
Siljan Ring 18, 171
Silurian 183
Simon, C. 56

Sinai Peninsula 127
Smith, R. I. L. 150
Sokoloff, W. 16
Sonora (Mexico) earthquake (1887) 60
Soter, S. (with Gold) 91
Source rock 168
South America 148, 157
South China Sea 73
South Sandwich Islands 131
Space dust 31
Spitsbergen 128
St. Helens, Mount 94
St. Lawrence River 47
Stahl, W. (with Tietze) 130
Stoqueler, 57
Streett, W. E. (with Gold) 136
Studier, M. H. (with Anders) 18, 34, 35
Sugisaki, R. 71
Sugiura, T. (with Sugisaki) 71
Sumatra 17, 128, 149, 150
Sungpan-Pingwu (China) earthquake (1976) 61, 62
Supercritical gases 140
Svalbard 128
Swabia earthquake (1911) 60
Sweden 16, 171
Swedish State Power Board 171, 172
Sylvester-Bradley, P. C. 20, 151

Tambora eruption (1815) 93, 98
Tanganyika, Lake 129
Tangshan earthquakes (1976) 67
Tashkent earthquake (1966) 71
Taupo, Lake 132
Teng, Ta-Liang (with Wallace) 62
Texas 113ff., 133
Tierra del Fuego 149
Tietze, K. 130
Tigris Valley 127
Tikhominrov, N. I. (with Beskrovny) 18

Titan 27
Toba eruption (75,000 years ago)
  93
Trofinov, M. V. (with Antesiforov)
  148
Tsunamis 49, 77

Udachnaya pipe 44
Ural Mountains 158
Uranium 148, 159
Uranus 22, 27
Urey, H. 35

Valdivia earthquake (1835) 59
Valyaev, B. M. 17, 181
Valyaev, B. M. (with Kropotkin) 19
Vanadium 143, 148
Velocity of propagation of seismic
  waves 72
Venezuela 158
Venus 22, 23, 27, 28, 29, 30
Vernadsky, V. I. 16
Vesuvius 51

Vlasov, K. A. 159
Volcan Fuego eruption (1974) 98
Voronoi, E. E. 19
Vosges (France) earthquake (1682)
  56
Voytov, G. I. (with Kravtsov) 43

Wakita H. 110
Wallace, R. E. 62
Warman, H. R. (with Kent) 126
Wehner, H. (with Tietze) 130
White, B. 164
Wilkening, L. 30
Williams, Samuel 53
Wittich, W. 68
Wyoming 157

Xenon 22

Yaroskevsky, A. A. (with Ronov)
  152
Yuma, Ft. (1852) 59